美国加州零排放车积分交易制度

中国清洁发展机制基金管理中心　编译

（2015 年）

中国商务出版社

图书在版编目（CIP）数据

美国加州零排放车积分交易制度／中国清洁发展机制
基金管理中心编译. —北京：中国商务出版社，2015.12
　ISBN 978 – 7 – 5103 – 1447 – 6

　Ⅰ. ①美…　Ⅱ. ①中…　Ⅲ. ①汽车排气 – 零排
放 – 排污交易 – 美国　Ⅳ. ①X511

中国版本图书馆 CIP 数据核字（2015）第 301926 号

美国加州零排放车积分交易制度

中国清洁发展机制基金管理中心　编译

出　版：中国商务出版社
发　行：北京中商图出版物发行有限责任公司
社　址：北京市东城区安定门外大街东后巷 28 号
邮　编：100710
电　话：010 – 64269744　64218072　（编辑一室）
　　　　010 – 64266119　（发行部）
　　　　010 – 64263201　（零售、邮购）
网　址：http：//www. cctpress. com
网　店：http：//cctpress. taobao. com
邮　箱：cctp@ cctpress. com　bjys@ cctpress. com
照　排：北京亮杰技贸有限公司
印　刷：北京九州迅驰传媒文化有限公司印装
开　本：787 毫米×1092 毫米　1/16
印　张：18　字　数：370 千字
版　次：2016 年 1 月第 1 版　2016 年 1 月第 1 次印刷
书　号：ISBN 978 – 7 – 5103 – 1447 – 6
定　价：73.00 元

序

生态文明建设关系人民福祉，关乎民族未来。加快推进生态文明建设是加快转变经济发展方式、提高发展质量和效益的内在要求，是积极应对气候变化、维护全球生态安全的重大举措。2015 年 6 月 30 日，中国向联合国气候变化框架公约秘书处提交了《强化应对气候变化行动——中国国家自主贡献》，明确提出二氧化碳排放到 2030 年左右达到峰值并争取尽早达峰，单位国内生产总值二氧化碳排放比 2005 年下降 60% 至 65%。

当前，我国经济发展已步入新常态，如何将资源环境制约挑战转化为机遇，调动各方力量低成本地实现绿色低碳发展转型，是急需解决的大课题。要构建科技含量高、资源消耗低、环境污染少的产业结构，加快推动生产方式绿色化，就必须转变观念，改革创新，充分发挥市场在资源配置中的决定作用，探索出一条在法制框架下、在政府引导下以市场减排和社会减排为主的发展路径。

汽车产业是我国经济发展和社会消费中的支柱性产业，它的发展对能源利用、碳减排、污染防治等工作有很大影响。学习借鉴国际成功经验，加快汽车产业的升级和转型，发展节能与新能源汽车非常紧迫。加利福尼亚州（下称"加州"）是美国汽车消费的第一大地区市场，也是汽车尾气排放污染最严重的州之一。从 20 世纪 90 年代开始，加州通过建立和实施零排放车积分交易制度（Zero Emission Vehicle Program），引导汽车生产和消费，有力促进了环境、产业和市场的协同良性发展，并迅速催生了世界级的新能源汽车企业特斯拉公司。这种基于市场导向的体制机制创新对于我国当前生态文明建设具有积极的实践借鉴意义。

加州零排放车积分交易制度的核心内容是以强制性政策创造了汽车生产企业的积分需求，催生了积分交易，形成了引导和驱动汽车产业升级与转型、鼓励低排放车和零排放车发展、减少环境污染和碳排放的市场机制。积分交易制度的原理与碳排放权配额交易同出一辙，具有"胡萝卜加大棒"双重功效，主要优势在于：第一，充分发挥了市场在资源配置中的决定性作用；第二，有效发挥了政府的政策制定和严格监管作用；第三，实施低成本减排，减少了对财政资金的依赖。

中国清洁发展机制基金（下称"清洁基金"）是国务院批准成立的、按照市场化模式管理的政策性基金。自 2007 年运行以来，清洁基金积极探索推广政府和社会资本合作（PPP）模式，较好地发挥了政府与市场、财政与金融、国内与国际的协同引导作用，撬动社会资金支持应对气候变化和绿色低碳发展事业的产业化、市场化、社会化和国际化发展。因清洁基金成功引入 PPP 模式促进低碳产业发展，2013 年，联合国气候变化框架公约秘书处和世界经济论坛向清洁基金联合颁发了"灯塔奖"，并推荐在其他发展中国家推广和复制。

为学习加州零排放车积分交易制度，2014 年初以来，清洁基金管理中心开展了专项调研，组织翻译了相关法规文件，帮助更多人员比较系统地了解积分交易制度的来龙去脉，近期又将这些工作成果梳理合编成本书，希望能为政府部门、产业机构、金融机构、研究和咨询机构以及相关从业人员提供有价值的参考，为我国节能和新能源汽车产业的发展加力。

财政部副部长 史耀斌

2015 年 11 月

译文说明

鉴于《加州零排放车法规》和《加州低排放车法规》英文原文专业技术性强，且用法律语言表述，为尽量保证忠于原文，本书第二部分的译文完全保持了原文的架构体系、行文格式和语言特点。为方便大家理解，特做以下几点说明：

一、加州法规编号

《加州法规》（*California Code of Regulations*）包含 4 级编号：主题（TITLE）、部门（DIVISION）、章（CHAPTER）、节（ARTICLE）。在第 13 主题（机动车相关内容）、部门编号为 3（加州空气资源委员会）、第 1 章（机动车污染控制装置）的第 3 节（新机动车污染控制装置的通过）下，17 个小节的内容构成了《加州低排放车法规》（*The California Low – Emission Vehicle Regulations*），《零排放车法规》是其中的一个组成部分①。

二、主要车型介绍

零排放车（ZEV）目前主要包括氢燃料电池车和纯电动车；升级版高技术部分零排放车（Enhanced AT PZEV）主要包括插电式混合动力车和氢内燃机车；高技术部分零排放车（AT PZEV）主要包括混合动力车，压缩天然气车以及甲醇燃料电池车；部分零排放车（PZEV）主要指非常清洁的传统车。

三、标题符号体系

英文原文标题编号顺序采用（a）、（1）、（A）、1、a、i 的六级编号体系。前五级标题为斜体，第六级为常规正体。

---

① 详见《加州法规》官方网站：https：//govt. westlaw. com/calregs/Browse/Home/California/CaliforniaCodeofRegulations？guid = I6AA005C02DDD11E197D9B83B68A61150&originationContext = documenttoc&transitionType = Default&contextData = （sc. Default）

四、英文缩写

为保持与英文原文一致以及译文简洁，本书保留英文原文中的缩写方式，但第一次出现时均提供中文译名全称。为方便查阅，有关章节均包含了关于缩写及其定义或解释的专门介绍。

五、技术文件引用

下列标注了节号的内容被法规直接引用作为技术支持：§1900《定义》（提供与《加州低排放车法规》有关的名词定义和解释）；§1961《尾气排放标准和试验程序——2004–2019 年款乘用车、轻型卡车和中型车辆》；§1961.1《温室气体尾气排放标准和试验程序——2009–2016 年款乘用车、轻型卡车和中型车辆》。

六、对"保留"的解释

《零排放车法规》仍在不断修订中，为此，法规条文中有意保留了标注为"Reserved"即"保留"的空白条款。译文保持了这一特色。

七、脚注说明

为帮助读者了解文中有关背景或专业解释，我们对一些内容加以脚注，并标示为"译者注"。

目　录

第一部分

美国加州零排放车积分交易制度解读

1

美国加州零排放车积分交易制度
介绍及对我国的借鉴意义

美国加州实施的零排放车积分交易制度作为一种具有政策强制性的市场机制，为在加州稳步扩大节能和新能源车市场份额，减少空气污染和碳排放，发挥着独特而有效的作用。

一、美国加州零排放车积分交易制度情况

（一）积分交易制度概况

为鼓励发展低排放车和零排放车，减少尾气排放和碳排放，加州环保署下属空气资源委员会（CARB）从 1990 年开始，以州法规方式，对乘用车、轻型卡车和中型车辆颁布尾气排放标准和测试程序，作为加州低排放车标准实施，并逐步纳入和加强零排放车的标准规定。2005 年开始实施零排放车积分交易，2008 年颁布的《加州零排放车法规》是其现行积分交易制度的依据，目前仍在修订完善中。

积分交易制度由积分目标、积分构成、积分交易、监督核查和惩罚措施等部分组成。其核心内容和最大特色是以强制性政策创造了车企积分需求，催生了车企间积分交易，形成了引导和驱动汽车产业转型升级的市场机制。积分交易制度的原理可类比碳排放权配额交易，具有"胡萝卜加大棒"的特征，设立过程体现了政府、汽车生产企业（下称车企）和社会公众的深度参与。

（二）积分交易制度主要内容

1. 积分目标和积分构成

在积分交易制度下，车企分为需获得积分和提供积分两大类。

在加州销售汽油车和柴油车等车型、拥有一定年销售规模的车企（下称 A 类车企），每年应实现相应的零排放车积分目标。

$$积分目标＝基数×目标系数$$

基数为车企的销售规模，目标系数是 CARB 规定的在基数中应实现的零排放车目标比例，体现政策导向、阶段性要求以及技术发展预期。例如，在 2013 年，CARB 设置的目标系数是 12%，如某车企当年销售规模是 5 000 辆，则它当年应完成 600 个积分。

生产零排放车型和部分零排放车型的车企（下称 B 类车企）提供积分。不同车型有不同积分，车型技术先进性和减排程度是主要评价因素。零排放车型包括纯电动车和氢燃料电池车，积分最高。部分零排放车型，如插电式混合电动车、混和动力电动车、非常清洁的传统车，技术先进性依次降低，可提供的积分也相应降低。每一车型的技术先进性随时间而变，相应积分将递减或停用。各车型积分在总积分中需满足一定比例，先进车型特别是零排放车型的积分占有比例逐渐增加。

另外，特定先进技术车型可在一定时期内获得额外积分鼓励，在交通系统中有连续使用记录的部分车型也可获额外积分。

2. 积分目标实现和积分交易

A 类车企有三种方案完成年度积分目标：一是自身生产零排放车和部分零排放车，自行获得积分；二是向 B 类车企购买积分；三是向州政府缴纳罚款。

A 类车企一般采用前两种方式，为维护企业形象，宁可多付钱买积分，也不选择交罚款，短期内更优先选择积分交易方式。迄今，加州尚无积分罚款记录。B 类车企自行决定是否出卖积分，政府不强制要求。交易价格由买卖双方决定，政府不干预。据了解，实际交易中，每个积分价格低者如 2000 美元，高者达 9000 美元。另外，因积分交易仅限于 A、B 两类车企之间，且存在不同车型积分随时间递减或停用的规定，所以尚未出现投机性交易。

CARB 建立"积分银行"电子信息管理系统，为 A、B 两类车企设立电子账户，记录积分交易和积分目标完成情况，并进行数据公开。

3. 监督核查

加州环保署是积分交易制度的主管部门，CARB 是执行管理机构，制定规则和监督核查。

各车企每年必须向 CARB 申请在加州销售汽车的车型注册证书，所售车辆必须满足加州尾气排放标准（高于美国联邦标准），2018 年后还须满足联邦温室气体管理计划规定。车企自己测量汽车尾气排放情况并登记，CARB 抽查真实性。

各车企每年自行向 CARB 申报积分目标，CARB 根据市场上独立公开的汽车销售数据，检查积分目标完成情况。

4. 惩罚措施

不遵守积分交易制度的车企须退出加州市场。连续两年不能完成积分目标的车企，按州立法，每个积分处以 5 000 美元罚款。

（三）积分交易制度实施情况

目前，加州有24家车企参加了积分交易制度，A类车企14家（6家大型车企和8家中型车企），B类车企10家，包括特斯拉（Tesla）。据了解，中国比亚迪汽车正在申请加入。

A类车企已明显感到积分交易制度压力，正在将其转化为转型升级的动力。B类企业获得了新的有利于加快发展的市场支持。

二、我国发展零排放车积分交易制度的意义和作用

（一）主动控制和减少汽车尾气排放

环境治理成本高，资金缺口大，财政负担不起以政府为主的治理模式。积分交易制度作为市场机制，面向控制排放总量，制定时间表，推动全产业链和消费者为减少汽车尾气排放采取行动，动员社会资金参与，让市场在资源配置中发挥决定性作用，是出路。

同时，积分交易制度是一种综合性政策体系，以积分交易为核心，还包括尾气排放标准、罚款（类似碳税）、燃油标准、对鼓励车型减免停车费等配套措施，是政府强制、市场驱动、社会经济运行三者的混合体，约束激励并重，尊重企业自主选择权，产生协同共治的整体效应，管得住、可执行。积分交易使政府奖罚措施显现化、货币化，提高了经济效率，降低了产业转型和减排的整体社会成本。

（二）促进汽车产业转型升级

环境治理表面是控污减排问题，本质是产业转型升级和新兴产业发展问题。积分交易制度不是着眼于提高燃油效率和燃油经济性的、服务于传统产业和产品改良的制度，而是把对尾气排放的控制从汽车个体层面上升到产业层面的制度，可以引发产业变革和推动创新，引导车企提高零排放车和低排放车的生产比重，降低传统汽油车和柴油车比重，从根本上促进汽车尾气减排，通过市场实现标本兼治。

（三）抓住新一代汽车产业发展的弯道超车历史机遇

受资源约束、价格上涨和环境保护等因素影响，高排放、低能效的汽油柴油车

已显颓势，低排放、高能效车是新一代汽车工业的发展方向，汽车产业竞争格局面临重新洗牌。在节能和新能源车的技术研发和商业化方面，我国与发达国家的差距不像在传统内燃机技术领域上那么大。积分交易制度将市场创新和政府支持结合在一起，有利于引导和推动汽车产业加快转型升级步伐，支持我国在新一轮汽车工业革命中，争取弯道超车机会。美国电动车企业特斯拉在加州积分交易制度支持下脱颖而出的经验，对我国有很大的启示意义。

2

美国加州从低排放车法规到零排放车法规的演进历程

汽车是现代社会的一种最常用交通工具，美国更被称为"汽车轮子上的国家"。汽车在为生活带来便利、为生产提供工具并成为一个重要的传统支柱产业的同时，排放的尾气也造成环境污染，其中温室气体排放还引发气候变化威胁。因此为控制和减少汽车尾气排放，亟需制定和实施越来越严格的标准。

在美国加州施行的机动车尾气排放标准，既有美国环保署制定的国家层面标准（以 Tire + 阿拉伯数字的形式标示排放级别），也有加州环保署制定的标准，即加州排放标准（以 LEV + 罗马数字形式标示排放级别）。加州环保署制定的机动车尾气排放标准总体上比美国环保署的标准更为严格，有效地引导了低排放车（Low Emission Vehicle，LEV）和零排放车（Zero Emission Vehicle，ZEV）的发展。1990 年以来，加州机动车尾气排放标准经历了从低排放车法规向零排放车法规发展的历程。这一进程基于技术发展并继续引导技术发展，推动了产业升级和替代。

一、加州低排放车法规的发展历程

加州环保署下属的加州空气资源委员会（CARB）负责制定涉及空气质量、汽车技术等多个方面的标准，以州法规的方式颁布和施行。1967 - 1989 年，CARB 主要通过制定空气质量标准和试图通过在机动车上增加减排装置来实现减排目的。1990 年后，CARB 主要围绕清洁燃油、低排放车和零排放车等整体概念，进行相关法规的设计和改进。

加州低排放车法规的发展可以分为三个阶段：第一阶段并行采用美国环保署的 Tier 1 和 LEV I 两种排放标准；第二阶段单独采用 LEV Ⅱ；第三阶段单独采用 LEV Ⅲ。

（一）第一阶段

为要求车企基于其车辆总产量中一定比例的车辆满足愈加严格的排放标准，

1990 年，加州环保署下属的加州空气资源委员会（CARB）通过了《加州低排放车法规》（第一阶段）（LEV I）为加州制定和发展有关控制和减少机动车尾气排放的法规迈出了第一步。作为第一阶段的法规，LEV I 内容相对简单，标准相对宽松。

LEV I 主要由三项法规组成：《1985 年及后续年款重型发动机和车辆的尾气排放标准和测试程序》、《1981—2006 年款乘用车、轻型卡车和中型车辆尾气排放标准和测试程序》和《1983 年及后续年款经联邦认证在加州销售的轻型摩托车认证程序》。在 LEV I 中，相同的气态污染物标准被应用于柴油和汽油动力车辆上，污染物排放量根据联邦测试程序 -75 来进行测量。重要的补充性联邦测试标准则在 2001—2005年期间分阶段引入。

（二）第二阶段

1998 年，CARB 通过了《加州低排放车法规》（第二阶段）（LEV Ⅱ），在 2004—2010 年间分阶段施行。LEV Ⅱ 主要由三项法规组成：《2004—2019 年款乘用车、轻型卡车和中型车辆尾气排放标准和测试程序》、《2009—2016 年款乘用车、轻型卡车和中型车辆温室气体排放标准和测试程序》和《机动车辆燃料挥发排放标准和测试程序》。LEV Ⅱ 规定，截至 2019 年款前，车企均可根据 LEV Ⅱ 为其车型进行认证：总重量在 8 500 磅以下的轻型卡车（LDT）及中型车辆（MDV）将被重新分类，并要求其必须满足乘用车（PC）的排放标准。大多数轻型货车以及运动型多功能车（在 LEV I 中被归为 MDV4 和 MDV5）也被要求必须满足 PC 排放标准。这些重新分类在 2007 年款之前分阶段完成。

LEV Ⅱ 中保留了 LEV、超低排放型车（ULEV）和特超低排放型车（SULEV），但是相应的标准比 LEV I 中更加严格。LEV Ⅱ 中新增加了关于部分零排放车（PZEV）的规定：除了要在排放污染物上满足特超低排放型车的标准外，还要在挥发性排放标准上满足零排放要求，且必须具有在 15 年内维持该排放性能的能力。

LEV Ⅱ 要求车企自 2010 年（LEV Ⅱ 施行的最后一年）起，每年都必须减少基于其车辆总产量的总排放量。对于轻型卡车，车企还必须满足越来越严格的非甲烷有机气体（NMOG）的车队平均排放标准。此处"车队"（fleet）是指某车型生产线生产的车辆数量。例如，2010 年 LEV Ⅱ 关于乘用车以及轻型卡车的车队平均非甲烷有机气体排放量为 0.035 克/英里；中型车辆虽然没有车队平均标准，但是车企却被要求认证一定比例的车辆满足排放标准，而且该比例逐渐增加。

所有类别车型的氮氧化物和颗粒物标准相对于 LEV I 标准都有所提高。所有车辆，无论使用什么种类的燃料，一律使用相同标准。在这种情况下，车辆只有安装先进的排放控制技术装置才能满足这样的要求。对柴油车而言，车辆通常会被要求安装颗粒物过滤器以及可以减少氮氧化物的催化剂装置。

除此之外，从 LEV Ⅱ 开始，法规融入了对减缓气候变化的考虑，新增了关于温室气体排放标准和测试程序的相关内容。

（三）第三阶段

《加州低排放车法规》（第三阶段）（LEV Ⅲ）于 2012 年 1 月通过，并于同年 12 月份进行了修订，在 2015－2025 年间分阶段施行。LEV Ⅲ 主要由三项法规组成：《2015 及后续年款乘用车、轻型卡车和中型车辆尾气排放标准和测试程序》、《2017 及后续年款乘用车、轻型卡车和中型车辆温室气体排放标准和测试程序》和《机动车辆燃料挥发排放标准和测试程序》。车企在 2015 年前可以自主选择对其车辆进行 LEV Ⅲ 认证，2020 年款以后所有车辆必须满足 LEV Ⅲ 标准。

在 LEV Ⅱ 的基础上，LEV Ⅲ 进行了以下改进：一是将原来分开的非甲烷有机气体和氮氧化物（NO_x）标准合并为非甲烷有机气体 + 氮氧化物（$NMOG + NO_x$）总量标准；二是对 2015－2025 年款的车辆设定更加严格的车队平均 $NMOG + NO_x$ 总量要求；三是新增了几类排放标准；四是对排放控制系统增加了耐用年限的要求。

LEV Ⅲ 规定要分阶段逐步施行更加严格的基于车辆总产量的新排放要求，轻型卡车总体的 $NMOG + NO_x$ 排放水平须在 2025 年达到与特超低型排放车辆相当的水平，即 0.030 克/英里。根据数据统计，三类主要车型即乘用车、轻型卡车以及中型车辆的平均 $NMOG + NO_x$ 排放量为 0.112 克/英里。因此，如果 LEV Ⅲ 所规定的平均排放标准可以实现，到 2025 年，这三类车型的排放量将会减少 73%。对中型车辆，虽然没有关于其车辆排放的总体平均要求，但是 LEV Ⅲ 却详细勾勒出了过渡路线图，它要求车企按照不断增加的比例来使其生产的中型车辆满足愈加严格的排放标准。

LEV Ⅲ 还规定，所有轻型卡车将会被要求满足一个日益严格的"零"挥发性排放标准，并且对检测所用的材料也要更加严格地选择。如果汽车可以装载臭氧污染物直接减排系统的话，例如，在车辆散热器上加装可以减少臭氧的催化剂装置，车企可以获得一个减排 NMOG 的积分。

二、零排放车法规的孕育和出台

加州政府通过在低排放车法规基础上进一步发展零排放车法规，使加州传统汽车产业转型升级已步入美国前列。这些法规不仅体现在 $NMOG + NO_x$ 标准等方面远高于《联邦清洁空气法案》要求，并且展现了雄心勃勃的温室气体减排目标，例如加州计划在 2020 年前将温室气体降低到 1990 年的水平，2050 年的温室气体排放在

1990 年水平的基础上减少 80% 等。CARB 认为,零排放车是加州实现自身温室气体减排目标、满足联邦空气质量标准、保护居民健康的一个必要选择。

CARB 在 1990 年出台的低排放车法规中早已包含了引导零排放车发展的思路,其中规定车企在满足 LEV 标准的同时,在其 1998 – 2000 年款的所售车辆中最少应该包含 2% 的零排放车,在其 2001 – 2002 年款中应最少包含 5% 的 ZEV,2003 年款及以后最少包含 10%。这是零排放车法规发展的雏形。由于当时电池成本过高、零排放车的车型较内燃机车车型较少,所以零排放车的发展未能按照预期进行。但是,关于零排放车发展的每两年回顾制度为其适时调整提供了机会。通过回顾,CARB 依据技术发展现状对设计零排放车发展的法规内容进行了修订,包括调整时间表、增加零排放车计划下可以获得积分的车型等。

在 1992 年和 1994 年的回顾中,CARB 确认零排放车技术应按照时间表发展,并重申了其对低排放车和零排放车的承诺。之后的审查将主要关注点都放在低排放车和零排放车相关技术上,并且专门成立了一个独立的电池技术咨询专家组来专门评估电池技术的发展。

在 1996 年的回顾中,CARB 根据 1995 年电池技术咨询专家组的报告和 CARB 成员的建议,投票废除了低排放车法规中对 1998 – 2001 年款的零排放车销售要求,但是依然维持对 2003 年款实现 10% 零排放车销售的要求。为了弥补因废除 1998 – 2001 年款零排放车销售要求而减少的减排量,CARB 和车企签署了一份备忘录,要求车企在全国范围内销售低排放车,并要求同 CARB 发展技术伙伴关系,部署开发先进动力电池的电动车。

在 1998 年的回顾中,CARB 将零排放车法规从低排放车法规中独立出来,为了增加法规的灵活性,还允许车企用部分零排放车(PZEV)的车辆来满足其零排放车的一部分销售要求,并且所获得的积分以 4∶1 的比例来抵消该车企应该产生的零排放车积分。大规模车企可以用部分零排放车来满足 60% 的销售要求,但是部分零排放车必须满足 LEV Ⅱ 所规定的超低排放车的排放标准、燃料挥发性排放标准,保证150 000 英里内其排放控制系统不会发生急剧老化问题,并满足第二代车载诊断系统的要求。

在 2000 年的回顾以及电池技术咨询专家组的建议导致了另一种技术类型车的出现——先进技术部分零排放车(AT – PZEV)。除了满足部分零排放车的要求外,这种车还采用了其他先进技术(电动驾驶系统、高压油气储存技术等)。

在 2001 年修订的低排放车法规中,要求根据零排放车单次充电的行驶里程适度增加其积分的乘数系数。这次修订维持了 2003 年款 10% 零排放车的销售要求,并要求以后逐年增加销售,到 2018 年将增至 18%。这次修订还要求从 2007 年开始,要将车企生产的运动型多功能车、皮卡等包括在基数内来计算其零排放车的销售要求。

2008 年，CARB 颁布了《加州零排放车法规》，主要包括三个部分：《2009 -2017 年款乘用车、轻型卡车和中型车辆的零排放车标准》、《2018 及后续年款乘用车、轻型卡车和中型车辆的零排放车标准》和《电动车充电要求》。

三、法规演进为产业和市场的发展发挥的约束与激励作用

从低排放车法规到零排放车法规，加州政府对机动车尾气排放标准的要求不断提高，实行愈加严格的管理制度。在低排放车法规发展阶段，CARB 主要致力于通过逐步改进传统内燃机车以减少尾气排放，而零排放车法规则通过改变车辆的动力源从而开辟车辆的零排放之路，并通过法规的方式将其与市场培育相结合，以市场为导向，从约束和激励两个方面发挥作用，在稳固低排放车发展的基础上，支持零排放车产业克服技术创新、技术应用和相关基础设施建设中可能遇到的困难，助其实现大规模的商业化，从而实现环境保护和应对气候变化目标。

加州机动车排放标准对车企首先是一种约束机制，基于环境容量限制要求车企所生产的车辆减少污染物排放和温室气体排放，逼迫车企改变原有发展模式，摆脱对传统技术路线的依赖和存量资产的拖累，为新一轮汽车动力技术革命成果让位。因此，在新一轮汽车动力技术革命中脱颖而出的不是通用、福特等传统汽车巨头，而是诞生于硅谷、没有传统技术禁锢的特斯拉公司。特斯拉公司的例子充分说明，约束高污染、高排放的传统产业可以为低污染、低排放的新兴市场的形成创造良好条件。

加州机动车尾气排放标准更是一种激励机制。它从低排放车起步，逐步加强对零排放车的发展激励，传递了明确的市场信号，并设计出零排放车积分交易制度，作为实施政策的市场机制措施。在稳固低排放车发展的基础上，促使车企加大对汽车零排放技术的研究投资，加快加州零排放车的市场培育。得益于这种"胡萝卜加大棒"的激励机制，自 1990 年以来，加州相关专利数量呈井喷式增长、高新技术溢出效应显著，零排放车市场规模不断扩大。著名车企尼桑公司首席执行官 Carlos Ghosn 曾经说："零排放车法规中对先进技术部分零排放车（AT-PZEV）的远景规划意味着混合动力电动车（HEV）大有商机，因为法规的强制性规定创造了一个巨大的零排放车市场。"

3

美国加州节能和新能源车发展动力分析

如今，以纯电动车为代表的加州新能源汽车市场发展方兴未艾，法律与政策、技术进步与创新、市场环境等因素共同驱动着节能和新能源车产业不断壮大。

一、法律与政策——产业发展的引导和保障

在加州机动车排放标准的演进历程中，我们可以清楚地观察到法律与政策在低排放车与零排放车产业发展和市场发展中所扮演的重要角色。

（一）宏观导向明确

鉴于零排放车具有的多种环境优势及其市场发展面临的挑战，布朗州长在 2013 年 3 月进一步发布了 B－16－2012 行政命令，在加州鼓励发展零排放车，从而保护环境、刺激经济增长、提高民众生活水平。州长行政命令设立了一系列长远目标和实现目标的路线图：到 2025 年，加州拥有的零排放车将到达 150 万辆；到 2050 年，加州由交通引起的温室气体排放在 1990 年的基础上将减少 80%。路线图表明了加州政府坚持发展零排放车、治理环境污染的信心，同时也向市场释放了明确的政策信号，有利于车企更加积极地参与到零排放车的研发、生产与推广进程中。

（二）微观规范严格

在制定宏观目标的同时，加州政府配套制定了详细、严格的行业标准，严格细化行业规范。从 1938－2011 年，联邦政府和加州政府层面通过的涉及到汽车方面的法律法规和相关标准达 80 多条，涵盖空气质量标准、燃料质量标准、内燃机燃烧效率以及汽车在运行或补充燃料时的排放等等各个方面，从而在很大程度上规范了有关汽车行业各方面技术的生产。这些微观层面的规范与低排放车法规和零排放车法规互为配套，相互支持。

（三）推进模式有序、目标设置可行

从低排放车法规到零排放车法规的历程说明，加州机动车排放标准在体现观念转变的同时，采取渐进式的推进模式，既保持一定压力，又尊重技术发展规律，使得零排放车的产业发展和市场推广避免出现概念炒作等不良现象，保证扎实推进。而且，为保证法规中目标可行、并非急功近利，所有目标的设定均以全面、具体的市场调查为依据，其中包括采取问卷、听证等多种方式，充分汇集车企、汽车销售商、专业咨询机构、非政府组织等利益相关方的意见。

例如：LEV Ⅲ规定了2021年之前，车企都必须按逐年上升的产量比例达到最终满足3毫克/英里的颗粒物排放标准，在2017年要求实现10%，2018年20%，2019年40%，2020年70%，直到2021年增至100%。

又如，为达到LEV Ⅲ规定的2025年目标，布朗州长在行政命令中设立了3个时间节点：2015年，加州大都市区的基础设施应可以适配零排放车；2020年，全加州基础设施应能够支持100万辆零排放车；2025年，加州将拥有150万辆零排放车，零排放车的市场占有率也将不断扩大。

二、技术进步与创新——产业发展的基础

技术进步和创新为从低排放车到零排放车的发展发挥了重要的支持作用。在加州低排放车法规到零排放车法规的演进过程中，技术和法律政策之间相互影响，技术的可行性和经济性是制定法律政策的基础，法律法规的日益严格也促进了技术的不断进步与创新。

（一）技术可行性、经济性——奠定法律政策的基础

低排放车法规和零排放车法规都建立在对已有技术基础的明确认识、能对未来技术发展理性分析和展望的基础之上。

以零排放车法规为例，技术基础大致分为两个层面：一是技术的可行性，二是技术的经济性。

技术可行性保证了所推广的零排放车车型已经在市场上存在，从技术研发角度说明零排放车技术是否可被开发并复制推广，以及在满足现有的技术标准后可否实现更严格同时又合理有效的新标准。例如，早在CARB准备制定零排放车法规之前，通用汽车公司就已经开发出了零排放车。CARB正是认识到了相关技术的可行性，才会认为在未来8－13年里零排放车有可能被推广，因此在起草《加州低排放车法

规》(第一阶段)(LEV Ⅰ)时,就在其中包含了涉及零排放车的条款,这成为孕育《加州零排放车法规》的起点。

技术可行性还表现在为复制推广程度较低但是较为先进的生产技术做出法律规定,有助于及时推广新技术,并推动技术创新沿着确定的方向进行下去。例如1961年,加州出现了美国第一个机动车排放控制技术——曲轴箱强制通风装置,与CARB关系紧密的加州机动车管理局即强制要求应用于在加州销售的车辆中。

然而,只有技术可行性不足以构成零排放车在加州全面推广的充分条件,只有降低技术成本,使得技术具备经济性,才能做到让零排放车普惠民众。技术的经济性是指所推广的零排放车车型的生产、销售、使用、维修等成本可以被控制在合理范围内。作为零排放车重要车型的纯电动车就曾经遭遇过因成本过高而难以在市场上推广的问题。1995年就有分析报告认为,因为所用电池造价过高,所以至少要到2000-2001年纯电动车才能投放市场。在2000年,CARB和车企的调查都显示,每辆纯电动车中的电池造价仍高达20 000美元以上,远远超出了普通消费者的承受水平,在当时阶段并不适宜大规模生产,只能小范围销售。因此,当时低排放车法规中所规定的纯电动车的生产比例就应该下降。随后,在法规修订时CARB进行了针对性的调整。

(二) 法律政策——推动技术发展

加州用立法形式颁布机动车排放标准,强化了对低排放车和零排放车技术开发、推广的推动作用。有研究表明,在颁布LEV Ⅰ和LEV Ⅱ的细化标准后,车企总是以比立法者所预期的速度更快地研发出价廉的新技术以使其产品满足相关标准。还有调查显示,一旦以立法形式对零排放车的某项标准或技术指标实施强制要求,相关技术创新就会争相迸发,来满足所针对的技术要求。

三、市场塑造——产业发展的环境

低排放车和零排放车的发展需要塑造一个新的市场。除了此前已经介绍的加州零排放车积分交易制度作为市场机制发挥作用外,还积极提升公众环保意识来孕育市场,建立法律体系来引导和规范市场,吸引私营部门资金来投入产业、壮大市场,让市场发挥资源配置的决定性作用。

(一) 提升公众环保意识孕育市场

1956年洛杉矶烟雾事件的爆发以及随之开展的一系列实验调查已经成为世界环

保历史中的标志性事件。环境恶化唤醒了人们的环保意识。可以看到，环保运动的不断发展使得加州政府不仅重视低排放车的推广，而且表现出符合技术发展趋势和市场发展趋势的前瞻性，在法律和政策方面主动经历了从低排放车法规到零排放车法规的演进。在这样一个背景下，加州机动车尾气排放标准所代表的环保要求已经成为加州车辆技术创新的出发点和归宿点。

加州消费者对汽车类型区分的要求是公众环保意识的增强的一个具体表现。为此，加州政府制定法规，要求不同环保程度的车辆必须加以标识，以方便消费者在选择交通工具时进行区分。例如：2007 年规定，车企必须在车辆上标示出其生产车型可能产生的致霾物和温室气体排放量，以帮助消费者明确该车对环境的影响；2008 年又规定，2009 年 1 月 1 日之后的所有新车上的环境标志，应为消费者提供一个比较车辆对气候变化和致霾物排放方面影响的工具。

（二）建立法律法规体系引导和规范市场

通过强有力的法律形式来强制实施技术标准、质量标准、提高各类低排放车型和零排放车型在所有新增车辆中的占比，引导和推动并举，也是加州政府的推进策略。技术指标类的法令为相关技术的发展和应用指明了方向，为企业融资和规模化生产指明了方向，成为了市场发展的一个重要风向标，发挥了引导作用。强制性的车型生产比例则为企业生产环保车型发挥了产业发展约束和推动作用。

在用法规形式来强制推广零排放车技术时，要特别注意不要将一定阶段内的汽车尾气减排目标设定得过高或过低。目标设低可能造成车企生产零排放车的潜力不能被充分动员，而高估技术创新的能力则可能会导致法规要求过于严格，致使政策无法落实，甚至遭致车企的共同反对。2001 年，加州零排放车车企和经销商曾起诉 CARB，当时过于严苛的零排放车标准正是此次官司的导火索。最后的解决方案表明，法律和政策的发展必须与当时实际技术发展状况保持一致，CARB 做出让步，提出了一种灵活的、满足零排放车的途径——"可选择满足途径（ACP）"。在 ACP 框架下，车企可选择利用零排放车、部分零排放车、先进技术部分零排放车、燃料电池车（FCV）之间的组合，来完成零排放车目标。上述调整还比过去增加了燃料电池车车型。

（三）吸引私人资金壮大市场

环境治理成本高，资金缺口大，以政府为主的治理模式会给财政带来极大的负担。选择具有杠杆效应的政策工具，用公共资金撬动社会资金，解决巨大的资金缺口问题，是加州走向零排放车的重要路径。公共资金主要流向市场的两个主体——

消费者、生产者，以采取补贴、低息贷款、合营入股等形式带动私人资金注入零排放车市场。

对于消费者，注入公共资金的主要目的是创造市场需求、引导消费者购买零排放车。2006 年，CARB 为购买或租借替代燃料和电动车辆的加州民众提供补贴 5 000 美元/年。2008 年，CARB 提供 480 万美元资金，用于支持清洗大约 42 万辆卡车和公共汽车的柴油发动机，帮助进行发动机翻新、更换以及装配其它节约燃料的装置；采取早期行动的企业将获得总额高达百万美元的资金援助，并且还能够获得低息贷款来运营清洁车辆。

对于生产者，加州为车企提供资金支持或补贴，还采取入股合营等方式，来激励和引导企业能够按照政府预期开展技术创新和生产活动。1999 年，加州成立了一家由车企、能源提供商、燃料电池车企以及加州政府共同组成的合资公司——加州燃料电池联盟，在加州开展燃料电池车示范。2001 年，CARB 加入了新成立的公私燃料电池协作机制，鼓励固定燃料电池商业化，从而减少对加州电网的需求。2008 年，CARB 的"卡尔摩耶"项目为加州若干空气质量管理区提供了 820 万美元资金，以项目的方式来促进更为清洁的柴油发动机的应用。这些措施都极大提高了零排放车生产链上各环节生产商的积极性，推动了产业的良性发展。

在零排放车市场上，加州对公共资金作用的定位是引导而非主导，公共资金的最重要目的是撬动社会资金，使得私人资本注入零排放车行业，在赚取利润回报的同时壮大零排放车市场。风险投资基金（VC）与私募股权投资基金（PE）在私人投资中扮演了重要角色。加州对环境保护的倡导让 VC 与 PE 敏锐地察觉到今后零排放车发展的潜力。以加州最重要的零排放车生产公司特斯拉为例，它从成立到 10 年后盈利，先后经历了多轮融资，总计近 7 亿美元。大量的风险投资不断地向新兴产业输送着活跃的资本，哺育了车企的成长。

（四）回顾修正——政企互动、适时调整

自零排放车法规颁布实施后，CARB 每两年都会对零排放车法规和低排放车法规的实施情况进行联合回顾。在回顾过程中，根据技术发展状况，零排放车的实施计划会对法规内容、时间框架做出调整。在 1995 年的回顾中，电池技术咨询专家组认为直至 2000 - 2001 年，氢燃料电池车才能将成本降低到符合消费者期望的水平并投入市场，这个评估结果比预定目标推迟了 3 年，CARB 也根据该评估结果适时调整了目标。

决策者据此制订的零排放车发展规划可以为车企提供明确的技术路线。2000 年的回顾显示，当时认为电池技术发展前景不容乐观，原本对 2003 年的成本估值在

12 500 美元的电池驱动纯电动车（BEV）现估值上升到了 20 000 美元以上。根据此项回顾，车企转而注重发展其他先进技术类型车辆，包括氢燃料电池车和迷你型电池车。一些电池生产企业，例如松下 EV 能源（宏达、丰田、福特纯电动车的主要供货商），开始转向于生产氢燃料电池，或者直接放弃继续生产纯电动车的电池。

高排放、低能效的汽油车和柴油车已显颓势，以混合动力车、混合燃料车、纯电动车和氢燃料电池车为代表的低排放、高能效车成为新一代汽车工业的发展方向，促进汽车产业转型升级和发展新兴产业迫在眉睫。零排放车法规不仅以法律形式明确规定了零排放车产业发展目标，也通过与企业的频繁互动，根据实际情况适时调整目标，使得决策者能够更清晰地掌握汽车工业的全局发展态势。

第二部分
美国加州零排放车和低排放车法规译文

1

加州零排放车法规

§1962.1 2009 – 2017 年款乘用车、轻型卡车和中型车辆的零排放车标准

（a）零排放车（ZEV）排放标准。如果车辆在任何或所有可能的行驶状态或条件下排放的任何标准污染物（或前体物）为零，执行官员将确认 2009 – 2017 年款新产乘用车（PC）、轻型卡车（LDT）和中型车辆（MDV）符合零排放车标准。

（b）ZEV 比例目标。

（1）ZEV 一般比例目标。

（A）基本目标。下表中列出了每家车企需达到的零排放车最低比例目标，即按照（b）(1)(C) 条款的要求并符合 1962.1（b）条款的规定，计算由车企生产制造并在加利福尼亚州（以下简称"加州"）交付销售的乘用车（PC）、一类轻型卡车（LDT1）和二类轻型卡车（LDT2）中必须为零排放车的比例。ZEV 目标将依据相应年款的非甲烷有机气体排放报告制定。

年款	最低 ZEV 要求
2009 – 2011 年	11 %
2012 – 2014 年	12 %
2015 – 2017 年	14 %

（B）计算符合 ZEV 比例要求的车辆数量。车企可以使用下述的三年平均法或当年预测法对 1962.1（b）(1) 条款中规定的 2009 – 2017 年款的 ZEV 目标进行计算。车企可以在年度基础上切换两种计算方法。年款平均产量仅限用于确定 1962.1（b）(1)(A) 条款对 ZEV 规定的目标，对于第 1900 章中车企规模的确定没有影响。在确定 ZEV 目标的过程中，某车企（例如车企 A）生产一辆 PC、LDT1 或 LDT2，但是由另外一家车企（例如车企 B）以其自身品牌在加州推广销售，则该车

应算在车企 B 名下。

1. 2009 – 2011 年款时间内，车企生产并在加州交付销售的 PC、LDT1 以及 LDT2（适用产量）按 2003 – 2005 年款车企生产并在加州交付销售的 PC、LDT1 以及 LDT2（适用产量）的三年平均产量计算。另一种代替上述三年平均产量的计算方法是，车企可以选择采用其同一年内生产并在加州交付销售的 PC、LDT1 以及 LDT2（适用产量）数量进行计算，以确定车企的 ZEV 目标。

2. 2012 – 2017 年款间，车企某年车辆的产量将按照车企先前第四、五、六年生产并在加州交付销售的 PC 和 LDT 三年的平均产量计算〔例如对于 2013 年款车辆，其 ZEV 目标将依据 2007 – 2009 年款的 PC 和 LDT 产量确定，对于 2014 年款车辆，其 ZEV 目标将依据 2008 – 2010 年款的 PC 和 LDT 产量确定〕。如果不使用上述计算以往年度产量的三年平均法，车企可以选择以年款当年生产并在加州交付销售的 PC 和 LDT 数量为基础履行 ZEV 目标。

（C）LDT2 渐进式 ZEV 目标。从 2009 年款设定 ZEV 标准开始，在确定 (b)(1)(A) 条款中规定的车企 ZEV 总目标时应考虑车企 LDT2 的产量，按下表所示，比例呈增加趋势。

2009	*2010*	*2011*	*2012 +*
51%	68%	85%	100%

（D）确定车企销售量时排除 ZEV。为符合 1962.1（b)(1)(B) 和 1962.1 (b)(1)(C) 条款规定，车企在计算其生产并在加州交付销售的 PC、LDT1 以及 LDT2 的数量时，车企应排除其自身生产并在加州交付销售的零排放车数量，以及该车企控股 50% 以上的附属公司生产并在加州交付销售的零排放车数量。

（2）大型车企的目标。

（A）大型车企至 2011 年款的主要目标。

2009 – 2011 年款间，车企必须至少使用 ZEV 或 ZEV 积分满足 22.5% 的 ZEV 目标，另外至少 22.5% 的目标需通过 ZEV，AT PZEV（先进技术部分零排放配额车，以下简称先进技术部分零排放车）或相应的车辆积分实现。车企剩余的 ZEV 目标可以通过 PZEV（部分配额零排放车，以下简称部分零排放车）或 PZEV 积分实现。

（B）大型车企 2011 年款车辆的可选目标。

1. III 型 ZEV 的最低产量。

a.〔保留〕。

b. 2009 – 2011 年款的目标。 2009 – 2011 年款间，选择履行可选目标的车

企必须获得相当于 2003 － 2005 年款期间生产、交付销售并投入使用的 PC、LDT1 和
LDT2（适用产量）年销售量 0.82% 的 ZEV 积分。ZEV 积分可以通过 2003 － 2005 年
款内生产、交付销售并投入使用的 ZEV 获取，不包括 NEV（低速电动车）① 或 0 型
ZEV，ZEV 积分按照下表每类 ZEV 车型与 III 型 ZEV 之间的积分替换率计算，或可
提交相应车辆产生的的等额积分。

ZEV 类型	与 III 型 ZEV 之间的积分替换率
I 型	2
1.5 型	1.6
II 型	1.33
IV 型	0.8
V 型	0.57

i. 如果 1997 － 2003 年款的 ZEV 符合 1962.1（f）条款的规定中关于延长服
务乘数的要求，并且该乘数的 33 年为 4 个 ZEV 积分，则车企可以在公历 2009 －
2011 任一年内使用该 ZEV 积分。

c. ［保留］。

d. ［保留］。

e. ［保留］。

f. 不包括交通系统的额外积分。

根据 1962.1（g）(5) 条款产生的交通系统的任何额外积分不能用于履行 1962.1
(b)(2)(B) 1.b 的目标。

g. 剩余积分的结转。 2005 － 2008 年款间，超额产量的 ZEV 积分可以结转
用于履行 1962.1 (b)(2)(B) 1.b. 条款规定的 2009 － 2011 年款最低产量目标，要
求结转的积分值应该基于积分使用当年的年款。自 2012 年款开始，结转积分将不能
用于履行 1962.1 (b)(2)(B) 1.b. 条款中规定的 ZEV 目标，但可用作 TZEV（过渡
型零排放车）、AT PZEV 或 PZEV 积分。

2009 － 2011 年款的 ZEV 积分可以结转两年用于履行 ZEV 目标。例如，
2010 年款的 ZEV 积分可以结转到 2012 年款，并仍然可以全额使用。自 2013 年款开
始，该积分只能用作 TZEV、AT PZEV 或 PZEV 积分，并且不能用于履行 ZEV 积分
目标，ZEV 积分目标只能使用 ZEV 产生的积分履行。

---

① 美国对 NEV 的定义是最高车速低于 45mph 的车辆。

h. ZEV 产量未达标的情况。车企针对 2009－2011 间任一年款选择 1962.1（b）（2）（B）条款中的可选目标后，于 2011 年款年底未能达到 1962.1（b）（2）（B）1. b. 条款中的目标要求的情况下，则必须在 2009－2011 年款间履行 1962.1（b）（2）（A）条款中的主要目标。

i. 取整约定。1962.1（b）（2）（B）1. b. 条款所规定的车企需要生产 ZEV 的数量应就近取整。

2. 履行 ZEV 的比例要求。2009－2011 年款间，某年内选择可选目标的车企，其 ZEV、AT PZEV 或 TZEV 或该类型车辆的积分应至少达到 45% 的 ZEV 目标。在任一年款内，用于履行可选目标的 ZEV 积分可以用于履行上述 45% 的目标，该目标可以通过生产 ZEV、AT PZEV 以及 TZEV 或该类车型的积分来履行，但 PZEV 除外。车企可以通过 PZEV 或其积分履行剩余的 ZEV 目标。

3. 2011 年款后可选目标取消。2011 年款后将不再使用 1962.1（b）（2）（B）条款中的可选目标。

（C）2009－2011 年款间大型车企对主要目标或可选目标的选择。如车企未于 2009 年款之前以书面形式告知执行官员其将于 2009 年款内选择可选目标，则车企应按主要目标履行 ZEV 达标任务。此后，车企须延续前一年选择履行目标的方式，除非于下一年款开始前以书面形式告知执行官员将选择可选目标。但是，在 2009－2011 年款间，如果车企在一年或多年内选择履行主要目标，可在 2011 年款年底前，出示在相应时间内达到 1962.1（b）（2）（B）1. b. 规定目标的证明，就可以选择 2009－2011 年款期间履行可选目标。

（D）2012－2017 年款间大型车企的目标。

1. 2012－2014 年款的目标。车企按照 1962.1（b）（1）（B）条款规定的任一种产量计算方法确定年度产量后，每年必须使用生产 ZEV 获得的积分履行 ZEV 目标，该积分至少占年度产量的 0.79%，且不能使用 NEV 或 0 型 ZEV 产生的积分，PZEV 积分可以履行目标的比例不得超过 50%，AT PZEV 积分履行的目标比例不得超过 75%。TZEV、0 型 ZEV 以及 NEV 积分履行的目标比例不得超过 93.4%（限制内容见 1962.1（g）（6）条款）。目标也可全部由生产 ZEV 获得的积分履行兑现。

2. 2015－2017 年款的目标。车企按照 1962.1（b）（1）（B）条款规定的任一种产量计算方法确定年度产量后，每年必须使用生产 ZEV 获得的积分履行 ZEV 目标，该积分至少占年度产量的 3%，且不能使用 NEV 或 0 型 ZEV 积分，PZEV 积分可以履行的目标比例不得超过 42.8%，AT PZEV 积分履行的目标比例不得超过 57.1%。TZEV、0 型 ZEV 以及 NEV 积分履行的目标比例不得超过 78.5%（限制内容见 1962.1（g）（6）条款）。目标也可全部由 ZEV 积分履行兑现。

3. 下表列举了车企在 2012 – 2017 年款间，履行 ZEV 目标及 TZEV、AT PZEV 以及 PZEV 种类占比最大时，车企应实现的比例目标。

年款	总 ZEV 比例目标	最低 ZEV 比例	TZEV、0 型或 NEV	AT PZEV	PZEV
2012 – 2014	12	0.79	2.21	3.0	6.0
2015 – 2017	14	3.0	3.0	2.0	6.0

4. 交通系统额外积分的使用。根据 1962.1（g）(5) 条款规定的交通系统内 ZEV 产生的任何额外积分可用于履行 ZEV 目标，履行比例最高为 10%，具体参见 1962.1（b）(2)（D）条款中的说明。

（E）〔保留〕。

（3）中型车企目标。2009 – 2017 年款间，中型车企可以通过 100% 使用 PZEV 或 PZEV 积分履行 ZEV 目标。2015 – 2017 年款间，中型车企总的积分比例目标为 12%。

（4）小型车企和自主小型车企的目标。小型车企和自主小型车企不需要履行 ZEV 目标比例。但是，小型车企和自主小型车企可以通过生产并在加州交付销售的 ZEV、TZEV、AT PZEV 或 PZEV 获得积分并卖给其他车企。

（5）〔保留〕。

（6）〔保留〕。

（7）小型车企、自主小型车企、中型车企的规模变化。

（A）加州产量增加。2009 – 2017 年款间，如果小型车企先前连续三年生产并在加州交付销售的新产 PC、LDT 以及 MDV 平均产量超出 4 500 辆，或者，如果自主小型车企先前连续三年生产并在加州交付销售的新产 PC、LDT 以及 MDV 平均产量超出 10 000 辆，这类车企将不再被视为小型车企或自主小型车企，并应开始履行中型车企的 ZEV 目标，计算始点为上一个连续三年之后的第六年。

如果中型车企先前连续三年生产并在加州交付销售的新产 PC、LDT 以及 MDV 平均产量超出 60 000 辆（即先前连续三年总产量超过 180 000 辆），这类车企将不再被视为中型车企，自上述连续三年之后的第六年或于 2018 年（以先出现的为准）开始，履行大型车企的 ZEV 目标。

如果车企在 2003 – 2017 年款内股权发生变化，致使该车企不再是中型车企，则须于第六年或 2018 年（以先出现的为准）开始履行新目标。

（B）加州产量减少。如果车企先前连续三年生产并在加州交付销售的新产 PC、LDT 以及 MDV 平均产量低于 4 500 辆、10 000 辆或 60 000 辆，这类车企应被视为小型车企、自主小型车企或中型车企，并应于下一年款开始履行小型车企、自主小型车企或中型车企的目标。

（C）变更所有权时加州产量的计算。某年款内，如果车企发生任何所有权变更，这将影响车企下一年款需履行的整体目标。如果车企在所有权变更时同时生产两种年款车型，则须使用较早年款确定下一年款的产量。在下一年款，小型车企、自主小型车企或中型车企的规模应依据其先前连续三年的平均产量判断。例如，如果 2010 年发生所有权变更，车企同时生产 2010 年款和 2011 年款车辆，导致需将车企 A 与车企 B 的产量相加，则 2011 年款车企 A 的规模将依据车企 A 和 B 在 2008 - 2010 年款的产量确定。在 2010 年款内，车企 A 的产量必须与车企 B、C 的产量合并，而且在此年款内所有权变更后不需要将车企 B 与车企 A 的产量合并，那么 2011 年款车企 A 的规模将依据车企 A 和 C 在 2008 - 2010 年款的产量确定。在任一情况下，1962.1（b）(7)（A）及（B）条款都将适用。

（c）部分零排放车（PZEV）。

（1）说明。本部分 1962.1（c）条款提供了识别在加州交付销售的车辆是否属于 PZEV 的标准。PZEV 不能认证为 ZEV，但可以获得至少 0.2 分的 PZEV 配额。

（2）基准 PZEV 配额。为了使车辆达到获取 PZEV 配额的标准，车企必须证明车辆符合以下所有要求。合格的车辆将得到 0.2 分的基准 PZEV 配额。

（A）SULEV（特超低排放车）标准。2009 - 2013 年款间，证明车辆符合 1961（a）(1) 条款规定的 15 万英里 SULEV 尾气排气标准。双燃料，灵活燃料及混合燃料车辆必须证明在使用两种燃料的情况下符合 15 万英里 SULEV 尾气排气标准。2014 - 2017 年款间，证明车辆符合 1961.2（a）(1) 条款规定的 15 万英里 SULEV 20 或 30 尾气排气标准，或 1961（a）(1) 规定的 15 万英里 SULEV 尾气排气标准。双燃料，灵活燃料及混合燃料车辆必须证明在使用两种燃料的情况下符合 15 万英里 SULEV 尾气排气标准。

（B）挥发性排放。2009 - 2013 年款间，证明车辆符合 1976（b）(1)（E）条款中的挥发性排放标准（零燃油挥发性排放标准）。2014 - 2017 年款间，证明车辆符合 1976（b）(1)（G）或 1976（b）(1)（E）条款中的挥发性排放标准。

（C）车载诊断系统（OBD）。证明车辆满足 1968.1 或 1968.2 中的 15 万英里（如果适用）车载诊断要求；并且

（D）延长的质保期。延长 2037（b）(2) 和 2038（b）(2) 条款中的性能和缺陷保修期至 15 年或 15 万英里（以先达到者为准）。对于使用零排放能源存储设备（如电池、超级电容或其他电能存储设备）作牵引动力的车辆，保修期为 10 年。

（3）零排放行驶里程（VMT）PZEV 配额。

（A）零排放行驶里程配额计算。符合 1962.1（c）（2）条款并有零排放行驶里程能力的车辆将会获得额外的零排放行驶里程 PZEV 配额，计算如下：

纯电续驶里程	零排放行驶里程 PZEV 配额
$EAER_u < 10$ 英里	0.0
$EAER_u \geqslant 10 - 40$ 英里	$EAER_u \times (1 - UF_{Rcda})/11.028$
$EAER_u > 40$ 英里	$3.627 \times (1 - UF_n)$ 其中， $n = 40 \times (R_{cda}/EAER_u)$

一辆车不能获得超过 1.39 分的零排放行驶里程 PZEV 配额。

《乘用车、轻型卡车和中型车辆 2009 - 2017 年款零排放车、混合动力电动车的加州尾气排放标准和测试程序》于 2008 年 12 月 17 日通过，上次修订是在 2012 年 12 月 6 日，城市等效纯电续驶里程（EAERu）和城市电量耗尽实际里程（Rcda）应分别根据其中的第 G.11.4 和 G.11.9 项内容确定。效用因子（UF）的确定应根据《2010 年 9 月 SAE 国际车辆表观信息报告 J2841》（2010 年 9 月修订），附录 B《车队效用因子》表格或使用表 2 效用因子方程系数中带有拟合系数的多项式曲线确定。

（B）可选程序。对于前一节 1962.1（c）（3）（A）中确定零排放行驶里程配额的可选方式，车企可以向执行官员申报使用可选程序以确定车辆零排放行驶里程的潜能，并作为总行驶里程的百分比，同时提交能够充分证明零排放行驶里程的工程评估。例如，可选程序可以规定零排放某种监管污染物（如氮氧化物）而非另一种污染物（如非甲烷有机气体）的车辆可获得 1.5 分的零排放行驶里程配额。

（4）先进 ZEV 零部件的 PZEV 配额。根据本条款，符合 1962.1（c）（2）要求的车辆有可能获得先进零部件 PZEV 配额。

（A）使用高压气体燃料或氢燃料存储系统。如果车辆配有高压气体燃料存储系统，其燃料储存能力为每平方英寸 3 600 磅或更高，并且车辆只通过该燃料运行，则车辆可获得 0.2 分先进零部件 PZEV 配额。如果车辆能够只通过存储于高压系统中氢燃料运行，此高压系统的储氢能力为每平方英寸 5 000 磅或更高，且燃料以非气态或在低温条件下存储，则车辆可获得 0.3 分先进零部件 PZEV 配额。

（B）使用符合条件的混合动力车（HEV）电驱动系统。

1. HEV 的分类。符合先进零部件 PZEV 配额的混合动力车基于下表中的标准可分为四种类型。

特性	D 型	E 型	F 型	G 型
电驱动系统峰值输出功率	≥10 kW	≥50 kW	零排放行驶里程配额；≥10英里 UDDS 工况下纯电续驶里程	零排放行驶里程配额；≥10英里 US06 工况下纯电续驶里程
牵引驱动系统电压	≥60 V	≥60 V	≥60 V	≥60 V
牵引驱动升压	是	是	是	是
回馈制动	是	是	是	是
怠速启停	是	是	是	是

2.〔保留〕。

3.〔保留〕。

4.〔保留〕。

5. D 型 HEV。如果车企证明其所展示的某 PZEV 符合 D 型 HEV 的所有标准，满足执行官员的要求，则该 PZEV 可获得 2009 – 2011 年款 0.4 分额外的先进零部件配额，2012 – 2014 年款 0.35 分的配额，2015 – 2017 年款 0.25 分的配额。

6. E 型 HEV。如果车企证明其所展示的某 PZEV 符合 E 型 HEV 的所有标准，满足执行官员的要求，则该 PZEV 可获得 2009 – 2011 年款 0.5 分额外的先进零部件配额，2012 – 2014 年款 0.45 分的配额，2015 – 2017 年款 0.35 分的配额。

7. F 型 HEV。如果车企证明其所展示的某 PZEV 符合 F 型 HEV 的所有标准，满足执行官员的要求，包括在 UDDS 工况下达到 10 英里或更高纯电续驶里程，PZEV 可获得 2009 – 2011 年款 0.72 分额外的先进零部件配额，2012 – 2014 年款 0.67 分的配额，2015 – 2017 年款 0.57 分的配额。

8. G 型 HEV。如果车企证明其所展示的某 PZEV 符合 G 型 HEV 的所有标准，满足执行官员的要求，包括在 US06 工况下达到 10 英里或更高纯电续驶里程，PZEV 可获得 2009 – 2011 年款 0.95 分额外的先进零部件配额，2012 – 2014 年款 0.9 分的配额，2015 – 2017 年款 0.8 分的配额。

9. 可分割性。如果 1962.1（c）(4)（B）1 – 8 的全部或部分被认为无效时，1962.1 的其余部分依然有效。

（5）低燃料周期排放的 PZEV 配额。专门使用具有低燃料周期排放的燃料的车辆，可获得 0.3 分的 PZEV 配额。为获得 PZEV 低燃料周期排放配额，车企必须使用同行评审的研究结果或其他相关信息向执行官员证明车辆燃料排放的非甲烷有机气体低于或等于 0.01 克/英里。燃料周期排放量必须根据近期的生产方法和基础设施设定进行计算，同时必须量化结果的不确定性。

（6）PZEV 配额的计算。

（A）车辆综合 PZEV 配额的计算。

在某年款内，合格车辆的综合 PZEV 配额为 1962.1（c）(6) 所列各项 PZEV 配额相加之和乘以 1962.1（c）(7) 所列的 PZEV 过渡性乘数，但上限不得高于 1962.1（c）(6)（B) 条款中规定的上限值。

1. 基准 PZEV 配额。 符合 1962.1（c）(2) 规定的标准，车辆获得 0.2 分的基准 PZEV 配额；

2. 零排放行驶里程 PZEV 配额。 如果存在零排放行驶里程 PZEV 配额，应按照 1962.1（c）(3) 条款确定；

3. 先进零部件 PZEV 配额。 如果存在先进零部件 PZEV 配额，应按照 1962.1（c）(4) 条款确定；

4. 燃料周期排放 PZEV 配额。 如果存在燃料周期排放 PZEV 配额，应按照 1962.1（c）(5) 条款确定；

（B）AT PZEV 配额上限。

1. 2009－2017 年款车辆的上限。 在乘以过渡性乘数之前 AT PZEV（包括基准 PZEV 配额）最多可以获得 3 分的配额。

2.〔保留〕。

（7）PZEV 乘数。

（A）〔保留〕。

（B）获得零排放行驶里程配额的 PZEV 引入过渡性乘数。 2009－2011 每一年款，根据 1962.1（c）(3) 条款获得零排放行驶里程配额的 PZEV，在加州出售或租赁 3 年或以上时间，且驾驶员在初次租赁期满时选择购买或者重新租赁 2 年或以上时间，那么该车过渡性乘数为 1.25。2011 年之后 1962.1（c）(7)（B) 条款不再提供乘数。

（d）零排放车乘数和积分条件。

（1）〔保留〕。

（2）〔保留〕。

（3）〔保留〕。

（4）〔保留〕。

（5）2009－2017 年款 ZEV 积分。

（A）ZEV 积分计算层级。 根据 ZEV 在下列 8 个不同 ZEV 层级的位置计算 ZEV 的积分：

ZEV 层级	UDDS 工况下 ZEV 续驶里程（英里）	快速充电能力
NEV	无最小值	不适用
0 型	< 50	不适用
I 型	≥50，<75	不适用
I. 5 型	≥75，<100	不适用
II 型	≥ 100	不适用
III 型	≥ 100	根据 1962.1（d）(5)（B）规定，必须能够在少于等于 10 分钟的充电时间内使得 ZEV 增加 UDDS 工况下 95 英里的续驶里程
	≥200	不适用
IV 型	≥200	根据 1962.1（d）(5)（B）规定，必须能够在少于等于 15 分钟的充电时间内使得 ZEV 增加 UDDS 工况下 190 英里的续驶里程
V 型	≥300	根据 1962.1（d）(5)（B）规定，必须能够在少于等于 15 分钟的充电时间内使得 ZEV 增加 UDDS 工况下 285 英里的续驶里程

1962.1（d）(5)（G）和（i）(10) 中定义了 I. 5x 型和 IIx 型车辆。

（B）快速充电。如果 III 型 ZEV 能够在 10 分钟或更短时间内使续驶里程达到 UDDS 工况下 95 英里，IV 型或 V 型 ZEV 能够在 15 分钟或更少的时间内使续驶里程分别达到 190 或 285 英里，则 2009 - 2017 年款的 III、IV 或 V 型零排放车被视为达到 1962（d）(5)（A）条款的快速充电能力要求。对于使用多种零排放车燃料的 ZEV，如插电式燃料电池车，执行官员可以不考虑 1962.1（d）(5)（B）条款快速充电的要求，可根据 UDDS 工况下 ZEV 续驶里程计算所得积分，详见 1962.1（d）(5)（A）条款的规定。

（C）2009 - 2017 年款 ZEV 积分。2009 - 2017 年款间，某 ZEV，包括 I. 5x 型和 IIx 型，不包 NEV 或 0 型，如在加州交付销售，可获得 1 个 ZEV 积分。根据其最早在加州投入运行的时间（不得早于该 ZEV 的年款），2009 - 2017 年款 ZEV 可获得额外积分。要想获得总积分，车辆必须在实施了《清洁空气法》第 177 章相关规定的州（以下简称"第 177 章所列各州"）或加州交付销售并投入运行。在车辆交付销售的州（加州或第 177 章规定的州）可以获得总积分。下表列出了一辆 ZEV 在 8 个不同层级中可获得的总积分，如果在某特定公历年或在该公历年结束后的 6 月 30 日之前投入运营，则总积分还包括不取决于投入运行的积分。如果在年款后第五个公历年的 12 月 31 日之后投入运行，则车辆不能获得积分。例如，如果车辆生产于 2012 年，在 2018 年 1 月 1 日投入运行，将无法获得 ZEV 积分。

根据 ZEV 类型以及生产、交付销售和投入运行确定的总积分		
层级	ZEV 投入运行的公历年份	
	2009 – 2011	*2012 – 2017*
NEV	0.30	0.30
0 型	1	1
I 型	2	2
I.5 型	2.5	2.5
I.5x 型	不适用	2.5
II 型	3	3
IIx 型	不适用	3
III 型	4	4
IV 型	5	5
V 型	7	2012 – 2014：7 2015 – 2017：9

（D）某些 ZEV 的乘数。2009 – 2011 年款 ZEV，不包括 NEV 或 0 型 ZEV，出售或租赁给驾驶员 3 年或以上时间，且驾驶员在初次租赁期满时选择购买或者重新租赁 2 年或以上时间，该车积分乘数应为 1.25。2011 年之后不再提供 1962.1（c）（5）（D）条款的乘数。

（E）计算在第 177 章所列各州及加州投入运行的特定 ZEV 数量。

1. 2009 年款规定。

a. 获得 ZEV 积分而非 NEV 及 0 型 ZEV 积分的大型车企和中型车企，如果其生产的 ZEV 被证明符合加州 ZEV 标准，或者在第 177 章所列各州被批准加入先进技术示范项目并投入运行，可以被认为符合 1962.1（b）条款的加州 ZEV 比例目标要求，其中包括 1962.1（b）（2）（B）条款的要求。

b. 获得 ZEV 积分而非 NEV 及 0 型 ZEV 积分的大型车企和中型车企，如果其所生产的 ZEV 被证明符合加州 ZEV 标准，或者在加州被批准加入先进技术示范项目并投入运行，可被算作符合第 177 章所列各州的零排放车比例目标，其中包括 1962.1（b）（2）（B）条款的要求。

2. 2010 – 2017 年款的规定。获得包括 1.5x 和 IIx 型 ZEV 车辆（除 NEV 及 0 型 ZEV）的 ZEV 积分的大型车企和中型车企，如果被证明符合加州零排放车标准，或者在加州或第 177 章所列各州被批准加入先进技术示范项目并投入运行，可以被视为履行 1962.1（b）条款所规定的加州或第 177 章所列各州的 ZEV 比例目标，条件是这里的积分要乘以一个比率，该比率是车企在接受积分的州中的适用产量（如 1962.1（b）（1）（B）条款的规定）与车企在加州的适用产量的比值（以下简称"比例值"）。第 177 章所列各州中产生的积分将基于比例值获取，而在加州根据 1962.1（d）（5）（C）条款中的规定可全值获取。但在 2010 至 2011 年款期间生产、交付销售以及投入运行的车辆或在加州作为先进技术示范项目以满足第 177 章所列各州执行 1962.1（b）（2）（B）条款的所得积分不需要与比例值相乘，不与比例值相乘的积分数量需要限定在 1962.1（b）（2）（B）1.b 条款规定的第 177 章所列各州的积分目标内。下表规定了符合第 177 章所列各州的 ZEV 类型。

车辆类型	年款
I、I.5 或 II 型 ZEV	2009 – 2017
III、IV 或 V 型 ZEV	2009 – 2017
I.5x 型或 IIx 型	2012 – 2017

3. 可选第 177 章中所列各州的达标途径。选用第 177 章所列各州的可选达标途径的大型和中型车企必须以书面形式通知执行官员以及每个第 177 章所列各州，时间不得晚于 2014 年 9 月 1 日。

a. 2016 至 2017 年款额外 ZEV 目标。选用第 177 章所列各州的可选达标途径的大型车企和中型车企，在 2012 至 2017 年款期间必须有额外的 ZEV 积分，包括不超过 50% 的 1.5x 型和 IIx 型 ZEV 积分，但不包括所有 NEV 和 0 型 ZEV 积分，在第 177 章所列各州内额外的积分数量等于下表中根据 1962.1（b）（1）（B）条款所确定的销售量比例：

年款	第 177 章所列各州的额外 ZEV 比例要求
2016	0.75%
2017	1.50%

对于用来履行车企在 2016 至 2017 年款间的额外 ZEV 积分目标的积分，1962.1（d）（5）（E）2 条款不适用。在 2016－2017 年款间，车企为履行 1962.1（d）（5）（E）3.a 的积分目标所生产的 ZEV 必须在第 177 章所列各州中投入使用，投入使用时间不得晚于 2018 年 6 月 30 日。

i. 在西部地区联营体和东部地区联营体内交易和转移 ZEV 积分。在西部地区联营体内，车企可以交易或转移特定年款 ZEV 积分，以履行 1962.1（d）（5）（E）3.c 条款中相同年款的目标，并且其积分值不会产生溢价。例如，某车企为弥补 2016 年款在 X 州的 100 个 ZEV 积分差额，该车企可以从西部地区联营体内的 Y 州转移出 100 个（2016 年款）ZEV 积分。在东部地区联营体内，车企可以交易或转移特定年款的 ZEV 积分，以履行 1962.1（d）（5）（E）3.c 中规定的相同年款的目标，并且其积分值不会产生溢价。例如，某车企为弥补 2016 年款在 W 州的 100 个 ZEV 积分差额，该车企可以从东部地区联营体内的 Z 州转移出 100 个（2016 年款）ZEV 积分。

ii. 在西部地区联营体和东部地区联营体之间交易和转移 ZEV 积分。车企可以在西部地区联营体和东部地区联营体之间交易或转移特定年款 ZEV 积分以履行 1962.1（d）（5）（E）3.c. 中规定的相同年款的目标。然而，在西部地区联营体和东部地区联营体之间，任何交易或转移的积分将产生其本身价值 30% 的溢价。例如，某车企为弥补在西部地区联营体中 2016 年款 100 个 ZEV 积分差额，该车企须从东部地区联营体中转移出 130 个（2016 年款）ZEV 积分。车企不得从加州 ZEV 银行向东、西部地区联营体中交易或转移积分，也不得从东、西部地区联营体向加利福尼亚 ZEV 银行中交易或转移积分。

b. TZEV 比例减少。如果大型和中型车企采取第 177 章所列各州的可选达标途径并完全符合 1962.1（d）（5）（E）3.a. 中规定的 2016－2017 年款额外 ZEV 要求，则在 2015 至 2017 年款中，车企可以根据 1962.1（b）（2）（D）2. 和 3. 的规定，在每个第 177 章所列各州允许的 TZEV 比例中减少一部分以满足 TZEV 比例。列举如下：

年款	2015	2016	2017
现有 TZEV 比例	3.00%	3.00%	3.00%
第 177 章所列州关于 TZEV 的可选达标途径的调整	75.00%	80.00%	85.00%
第 177 章所列州的可选达标途径的新 TZEV 比例	2.25%	2.40%	2.55%

车企可用 ZEV 积分或 TZEV 积分来达到上述减少的 TZEV 比例。减少的 TZEV 比例同时也降低了 1962.1（d）（5）（E）3. c. 中规定的总的 ZEV 百分比要求。

i. 在西部地区联营体和东部地区联营体内交易和转移 TZEV 积分。在西部地区联营体内，从 2015 年款开始，车企可以交易或转移特定的 TZEV 积分用于满足 1962.1（d）（5）（E）3. c. 条款中同一年款的目标，并且积分值不产生任何溢价。例如，某车企为弥补 2016 年款在 X 州的 100 积分差额，该车企可以从西部地区联营体内的 Y 州转移出 100（2016 年款）TZEV 积分。在东部地区联营体内，车企可以交易或转移部分 TZEV 积分来满足 1962.1（d）（5）（E）3. c. 条款中同一年款的目标，并且积分值不产生任何溢价。例如，某车企为弥补 2016 年款在 W 州的 100 积分差额，该车企可以从东部地区联营体内的 Z 州转移出 100（2016 年款）TZEV 积分。

ii. 在西部地区联营体和东部地区联营体之间交易和转移 TZEV 积分。车企可以在西部地区联营体和东部地区联营体之间交易或转移特定的 TZEV 积分来履行 1962.1（d）（5）（E）3. c. 条款设定的相同年款的目标。但任何交易或转移的积分将产生其本身价值 30% 的溢价。例如，某车企为弥补 2016 年款在西部地区联营体中的 100 个积分差额，该车企可以从东部地区联营体中转移出 130 个（2016 年款）TZEV 积分。车企不得从加州 ZEV 银行向东、西部地区联营体交易或转移积分，或者是从东、西部地区联营体向加州 ZEV 银行交易或转移积分。

c. 总的目标比例。在第 177 章所列各州，最低 ZEV 目标以及 1962.1（b）条款中有关 AT PZEV 及 PZEV 的允许比例，对选择第 177 章所列州的可选达标途径的大型车企和中型车企仍然有效。然而第 177 章所列各州的可选途径要求车企达到额外 ZEV 目标，并允许车企达到减少的 TZEV 比例，如上述 1962.1（d）（5）（E）3. a. 和 b. 条款中的说明。如果车企选择了可选达标途径，并获得用于履行 ZEV 最低目标的最少 ZEV 积分，同时 TZEV、AT PZEV 及 PZEV 积分所占比例最大，在此情况下，在 2015 – 2017 年款期间，第 177 章所列各州总的年度比例目标如下表所示：

年份	可选途径的总 ZEV 比例	可选途径的最小 ZEV 积分	可选路径的 TZEV 积分	AT PZEV 积分（不变）	PZEV 积分（不变）
2015	13. 25%	3. 00%	2. 25%	2. 00%	6. 00%
2016	14. 15%	3. 75%	2. 40%	2. 00%	6. 00%
2017	15. 05%	4. 50%	2. 55%	2. 00%	6. 00%

d. 报告要求。以年为基础，截至年款结束后的公历年 5 月 1 日，选择第 177 章所列各州可选达标途径的车企应当以书面形式向执行官员，以及第 177 章所列各州提交报告，包括详细项目单，说明车企已经在第 177 章所列各州以及东、西部地区联营体中履行了 1962.1（d）(5)(E) 3. 条款的目标。详细项目单应包括以下内容：

  i. 车企在第 177 章所列各州交付销售的 PC 和 LDT 的总适用产量，如 1962.1（b）(1)(B). 中规定。

  ii. 构造、型号、车辆识别码、获取的积分以及车企为履行 1962.1（d）(5)(E) 3. a. 条款的额外 ZEV 目标而在第 177 章所列各州交付销售并投入使用 ZEV 的州名。

  iii. 构造、型号、车辆识别码、获取的积分以及车企为履行 1962.1（d）(5)(E) 3. c. 条款目标而在第 177 章所列各州交付销售 TZEV 的州名，以及交付销售并投入使用 ZEV 的州名。

  e. 未能达到第 177 章所列各州的可选途径的目标。在相应地区联营体内第 177 章所列各州，选择第 177 章所列各州的可选达标途径的车企，截至 2018 年 6 月 30 日未能履行 1962.1（d）(5)(E) 3. a. 条款的目标，则从 2015 – 2017 年款，应遵守 1962.1（b）条款中规定的总 ZEV 比例目标，此外，1962.1（d）(5)(E) 3. a. 条款中规定不再适用。ZEV 积分在第 177 章所列各州之间的转移无效，ZEV 积分将退回到获取积分所在州。采取第 177 章所列各州的可选达标途径的车企，在某个年款或在相应联营范围内第 177 章所列各州中，如果未履行 1962.1（d）(5)(E) 3. b. 条款的目标，或根据 1962.1（g）(7)(A) 条款在特定时间内、使用特定积分补足所欠积分，则在 2015 – 2017 年款中，应遵守 1962.1（b）条款中的总 ZEV 比例目标，1962.1（d）(5)(E) 3. b. 条款中规定的联营规定不适用。如果车企未能履行这些目标，任何 TZEV 积分在第 177 章所列各州之间的转移都是无效的，同时 TZEV 积分将退回到获取积分所在州。车企如果未能在 2018 年款年末之前补足 ZEV 所欠积分，其罚款数额将在第 177 章所列各州分别计算。

  f. 1962.1 条款中的规定适用于选择第 177 章所列各州的可选达标途径的车企，但 1962.1（d）(5)(E) 3 条款特别修改部分除外。

  （F）NEVs。从 2010 年款开始，为获得 1962.1（d）(5)(C) 条款规定的积分数量，NEV 必须符合 1962.1（d）(5)(F) 条款中的以下规格和要求。

  1. 技术参数。2010 – 2017 年款间，当符合下列全部技术参数时，NEV 可获得积分：

  a. 加速度。在 332 磅有效载荷、电池在 50% 荷电状态下启动时，车辆在 6

秒或更短时间内完成 0 - 20 英里每小时加速。

b. 最高速度。在 332 磅有效载荷、电池在 50% 荷电状态下启动时，车辆的最高速度下限为 20 英里每小时。按照 49 CFR 571. 500（68 FR 43972，2003 年 7 月 25 日）① 进行测试时，车辆最高速度不应超过 25 英里每小时。

c. 等速续驶里程。在 332 磅有效载荷、启动时电池处于 100% 荷电状态下，并以最高速度等速运行时，最低续驶里程为 25 英里。

2. 电池要求。2010 - 2017 年款 NEV 必须配备一个或多个密封、免维护电池。

3. 保修要求。2010 - 2017 年款的 NEV 动力系统，包括电池组在内，保修期必须至少为 24 个月。NEV 保修期的前个六月内必须完全保修，剩余的保修期可选择延长保修（可购买）以及按比例进行保修。如果按比例延长保修，保修或退还的电池组初始价值的百分比至少不低于剩余的按比例分配的保修期比例。为了便于进行此类计算，电池组的寿命必须以时间间隔表示，单个时间间隔不大于三个月。除此之外，车企可以选择在整个延长的保修期限内负责电池组 50% 初始价值的保修服务。

4. 批准配额之前，执行官员可以要求车企提供具有代表性的车辆和电池保修单的副本。

5. NEV 充电要求。2014 - 2017 年款 NEV 必须满足 1962. 3（c）（2）条款中规定的充电连接标准。

（G）1.5x 型和 IIx 型汽车。从 2012 年款开始，为获得 1962. 1（d）（5）（C）条款中的积分数量，1.5x 型和 IIx 型车辆必须符合以下技术参数和要求：

1. PZEV 要求。1.5x 型和 IIx 型车辆必须符合 1962. 1（c）（2）（A）至（D）条款中规定的全部 PZEV 目标。

2. G 型 ZEV 要求。1.5x 型和 IIx 型车辆必须符合 1962. 1（c）（4）（B）条款中规定的 G 型先进零部件配额的要求。

3. 辅助动力单元。车辆辅助动力单元首次启动并进入"电量维持混合运行"后，UDDS 工况下的续驶里程必须小于或等于车辆辅助动力单元启动前的 UDDS 工况下的纯电续驶里程。车辆的辅助动力单元不能在用户可选驱动模式下启动，除非用于牵引的能量储存系统完全耗尽。

4. 最小的零排放里程要求。

---

① 译者注：联邦法规（美国）第 49 部分 交通 第 571 章 联邦机动车安全标准（68 卷，43972 页）

车辆类型	UDDS 工况下零排放里程
I.5x 型	≥75 英里，<100 英里
IIx 型	≥100 英里

(e) ［保留］。

(f) 大于等于 10 英里零排放里程的 1997 - 2003 年款的 ZEV 和 PZEV 的延长服务乘数。 除 NEV 之外，如果车企在 2009 至 2011 年间生产制造 1997 - 2003 年款零排放里程至少 10 公里的 ZEV 或 PZEV，并且每年都进行注册以在加州公共道路上行驶，那么车企在车辆最初三年运行后就可以获得额外 ZEV 或者 PZEV 乘数。对于 2003 年 4 月 24 日之前开始的额外运行年份，如果车辆在当年被新租用或出售，则车企将获取 0.1 倍的 ZEV 积分；该积分，包括乘数，都是在车辆最初开始运行后的第四年起逐年计算。对于 2003 年 4 月 24 日或之后开始的额外运行年份，如果车辆在当年被新租用或出售，则车企将获取 0.2 倍的 ZEV 积分；该积分，包括乘数，都是在车辆最初开始运行后的第四年起逐年计算。在每个连续的运行年份之后，车企出具报告并获取延长服务乘数。2011 年款后不能获取额外积分。

(g) 积分的产生和使用；处罚的计算。

(1) 说明。 生产制造并在加州交付销售 ZEV 或 PZEV 的车企，如果在某年款超过 1962.1（b）规定的 ZEV 目标，该车企将根据 1962.1（g）的规定获得积分。

(2) 积分计算。

(A) ZEV 积分。 2009 - 2014 年款间，车企在某年款获取的 ZEV 积分数量应该以克每英里的非甲烷有机气体（g/mi NMOG）为单位来表示，该值等于车企生产并在加州交付销售的总 ZEV 获得的积分减去车企为满足该年款 ZEV 目标，而生产并在加州交付销售的 ZEV 获取的积分数，然后乘以 2009 - 2011 年款 PC、LDT1 或 LDT2 或者 2012 - 2014 年款 PC 和 LDT1 的 NMOG 平均目标。

2015 - 2017 年款间，车企在某年款获取的 ZEV 积分数量应以积分为单位来表示，并且等于车企为满足 ZEV 目标而生产制造并在加州交付销售的 ZEV 获得的积分数，或者是，在适用的情况下，1962.1（d）(5)(E) 3 条款中的积分目标，并将这些积分从车企生产并在加州交付销售的总 ZEV 积分中减去。

(B) PZEV 积分。 2009 - 2014 年款间，车企在某年款获取的 PZEV 积分数量应以克每英里的 NMOG 为单位来表示，并且等于车企生产并在加州交付销售的 PZEV 获得的 PZEV 配额，减去车企为履行 ZEV 目标而生产并在加州交付销售的 PZEV 获取的积分数，然后乘以 2009 - 2011 年款 PC、LDT1 或 LDT2 或者 2012 - 2014 年款 PC 和 LDT1 的 NMOG 平均目标。

2015 - 2017 年款，车企在某年款获取的 PZEV 积分数量应以积分为单位来表示，并且等于为履行 ZEV 目标而生产及在加州交付销售的 PZEV 获得的积分数，或者是，在适用的情况下，1962.1（d）（5）（E）3 条款规定的目标，并将其从车企生产并在加州交付销售的 PZEV 积分数中减去。

（C）独立的积分账户。车企的 ［i］ZEV、［ii］I.5x 型和 IIx 型车辆、［iii］TZEV、［iv］AT PZEV 和 ［v］其他所有 PZEV 以及 ［vi］低速电动车的积分应该分别维护。

（D）舍入积分。对于 2012 - 2014 年款，只在最后的积分和借记总额使用惯例的舍入方法，ZEV 积分和借记应被舍入到 1/1 000 位。对于 2015 - 2017 年款，只在最后的积分和借记总额使用传统的舍入方法，ZEV 积分和借记被舍入到 1/100 位。

（E）转换 NMOG ZEV 积分（单位：克/英里）为 ZEV 积分。2014 年款目标履行完成后，所有车企 ZEV、I.5x 型和 IIx 型、TZEV、AT PZEV、以及 NEV 的账户将由 NMOG ZEV 积分转换为积分。每个 NMOG 账户余额将除以 0.035。从 2015 年款起，积分不再用克/英里为单位表示，而只表示为积分。

（F）2017 年款后转换 PZEV 和 AT PZEV 积分。2017 年款后，车企 PZEV 和 AT PZEV 积分账户可在转换后用于履行 1962.2（b）规定的目标。对于大型车企，PZEV 积分账户将被折价为 93.25%，AT PZEV 积分账户将被折价为 75%。对于中型车企，PZEV 和 AT PZEV 的积分账户将被折价为 75%。将在 2017 年款目标履行完成后一次性计算。

（3）MDV 和 LDT 的 ZEV 积分，不包括 LDT1。被车企归类为 MDVs 或 LDTs 而不是 LDT1 的 ZEV 和 PZEV 可计入 PC、LDT1 和 LDT2（适用的）的 ZEV 目标，并且如果车企指定，这些 ZEV 和 PZEV 包含在 ZEV 积分的计算当中，如 1962.1（g）的规定。

（4）先进技术示范项目的 ZEV 积分。

（A）TZEV。2009 - 2014 年款 TZEV，如果其在两年或更长的时间内被列为加州先进技术示范项目中，即使其没有"交付销售"或在加州机动车管理局登记，仍可获取 ZEV 积分。要取得这样的积分，车企必须向执行官员证明，车辆会经常使用，可以评估安全事宜、基础设施、燃料规格或公众教育等相关的问题，并且在最初运行的两年，50% 及以上的时间将在加州运行。上述情况中的车辆有资格获得与投入使用时得到的配额与积分相同的积分数额。要决定车辆积分，示范车型的年款需与相同时间内指定传统车型的年款相一致。根据 1962.1（g）（4）条款，车企每年在第 177 章所列各州每个车型最多有 25 辆车获取积分。超过 25 辆上限的车辆将没有资格获得先进技术示范项目积分。

（B）ZEV。2009 – 2017 年款 ZEV，包括 I.5x 型和 IIx 型车辆，不包括 NEV 以及 0 型 ZEV，有两年或更长的时间被列为加州先进技术示范项目，即使其没有"交付出售"或在加州机动车管理局登记，仍可获取 ZEV 积分。要取得这样的积分，车企必须向执行官员证明，车辆会经常使用，可以评估安全事宜、基础设施、燃料规格或公众教育相关的问题，在最初运行的两年，50% 及以上的时间车辆将在加州运行。这种车辆有资格获取与投入使用时获得的配额与积分相同的积分数额。要决定车辆积分，示范车型的年款需与相同时间内指定传统车型的年款相一致。根据 1962.1（g）（4），车企每年在第 177 章所列各州，每个车型最多有 25 辆车获取积分。超过 25 辆上限的车辆将没有资格获得先进技术示范项目积分。

（5）交通系统的 ZEV 积分。

（A）概述。2009 – 2011 年款 ZEV，如其在交通系统中运行连续两年或更长的时间，则该车辆可以获得额外的 ZEV 积分，除非在下述子项（g）（5）（C）中做出了其它规定，否则额外获得的这些积分可以按照和该类别车辆获得的其它积分相同的使用方式进行使用。针对 2012 – 2017 年款，如 ZEV、I.5x 和 IIx 型车辆或 TZEV 在交通系统中运行连续两年或更长时间，则该车辆可以获得额外 ZEV 积分，除非在下述子项（d）（5）（E）2 和（g）（5）（C）中做出了其它规定，否则额外获得的这些积分可以按照和该类别车辆获得的其它积分的相同使用方式进行使用。如果 2009 至 2011 年款 AT PZEV 或 PZEV 在交通系统中运行，则该车辆可以获得额外的 ZEV 积分，除非在下述子项（g）（5）（C）中做出了其它规定，否则额外获得的这些积分可以按照和该类别车辆获得的其它积分的相同使用方式进行使用。低速电动车不能获得针对交通系统的积分。要取得积分，车企必须满足执行官员的要求，向其证明，其制造的该型车辆将成为采用创新交通系统项目的一部分，子项（g）（5）（B）中规定了上述创新交通系统的信息。

（B）获得的积分。根据（g）（5）中的规定，如想获得额外的积分，则该项目至少必须证明：[i] 共同使用 ZEV、I.5x 和 IIx 型车辆、TZEV、AT PZEV 或者 PZEV；同时，[ii] 采用了诸如预约管理、一卡通系统、仓库管理、位置管理、收费开票和实时无线信息系统等在内的"智能"新技术。如果某个项目，除了满足上文中的 [i] 和 [ii] 之外，还具有联动运输功能，则该项目可获得更多的额外积分。对于 ZEV 来讲（不包括 NEV），如果某个项目具有联动运输功能，如专用停车场及充电设施等，但是不能满足上文中的 [i] 和 [ii]，则该项目仍可获得更多的额外积分。每辆车能够获得的最多积分由执行官员根据车企或项目经理（在适当情况下）提交的申请进行决定。每辆车获得的积分最多不应超过下文中给出的规定数量。

车型	年款	共同使用，智能	联动运输
PZEV	截至 2011 年	2	1
AT PZEV	截至 2011 年	4	2
TZEV	2009 – 2011 年	4	2
ZEV	2009 – 2011 年	6	3
TZEV	2012 – 2017 年	0.5	0.5
ZEV 和 I. 5x 和 IIx 型 ZEV	2012 – 2017 年	0.75	0.75

（C）交通系统积分的使用上限。

1. ZEV。根据（g)(5) 条款的规定，ZEV、I. 5x 型和 IIx 型车辆获得或分配的积分，不包括车辆自身获取的所有积分，最高可抵偿该车企应履行的某年款 ZEV 目标的十分之一，或者可用于抵偿该车企必须用 ZEV 履行的 ZEV 目标的十分之一，请参见 1962. 1（b)(2)(D) 3. 条款中的规定。

2. TZEV。根据（g)(5) 条款的规定，TZEV 获得或分配的积分，不包括车辆自身获取的所有积分，最高可抵偿该车企应履行的某年款 ZEV 目标的十分之一，或者，如情况适用的话，最高可抵偿该车企根据 1962. 1（d)(5)(E) 3. 条款规定的总 ZEV 目标的十分之一。但上述积分只能按照和该类别车辆获取的其他积分的相同使用方式进行使用。

3. AT PZEV。根据（g)(5) 条款的规定，AT PZEV 获得或分配的积分，不包括车辆自身获取的所有积分，在某年款中最高可抵偿该车企应履行的 ZEV 目标的二十分之一，但上述积分只能按照和该类别车辆获取的其他积分相同的使用方式进行使用。

4. PZEV。根据（g)(5) 条款的规定，PZEV 获得或分配的积分，不包括车辆自身获取的所有积分，在任一特定的年款中，最高可抵偿该车企应履行 ZEV 目标的五十分之一，但上述积分只能按照和该类别的车辆获取的其他积分相同的方式进行使用。

（D）交通系统积分的分配情况。如果某车辆已经按照 1962. 1（g)(5)(A) 条款中的时间段应用到某个项目中，则执行官员会将相应的积分分配给项目经理，或者在没有单独项目经理的情况中，分配给车企。执行官员根据由项目经理提交的并由参加该项目的所有车企签署的书面推荐，将对应的积分分配给车企；上述积分不必严格按照已投入运行车辆的数目进行分配。对于在 2017 年款后推出的用于交通系统的车辆，将不再给其分配此类积分。

（6）ZEV 积分的使用。2009 – 2014 年款，车企通过向执行官员提交符合

1962.1（b）条款规定的克/英里 ZEV 积分，以满足特定年款中该车企的 ZEV 目标。对于 2015 – 2017 年款，车企通过向执行官员提交符合 1962.1（b）条款规定的 ZEV 积分，以满足特定年款中该车企的 ZEV 目标。每个类别的积分可以用来满足针对该类别的目标，也可以用来满足比该类别获取 ZEV 积分更少的那些类别的目标，但是不能用于满足比该类别获取 ZEV 积分更多的那些类别的目标。例如，TZEV 产生的积分可用于满足 AT PZEV 的目标，但不能用于满足那些必须用 ZEV 积分才能履行的目标。这些积分可以由车企预先获取或由该车企从另一方处获得。

（A）NEV。2001 – 2005 年款间，交付销售并投入运行的 NEV 所获得的积分最多只能用来满足下表中的比例限制：

年款	ZEV 目标：	完成每项目标的NEV 比例限制[1]：
2009 – 2011	必须用 ZEV 完成	50%
2009	可以用 AT PZEV 而非 PZEV 完成	75%
2010 – 2011		50%
2009 – 2011	可以用 PZEV 完成	无限制
2012 – 2017	必须用 ZEV 完成	0%
	可以用 TZEV 和 AT PZEV 完成	50%
	可以用 PZEV 完成	无限制

1. 本表中的目标为 1962.1（d）(5)（E）3. 条款中规定的目标。

此外，2006 – 2017 年款，投入运行的 NEV 获取的积分可用于满足下表中描述的比例限制：

年款	ZEV 目标：	完成每项目标的NEV 比例限制[1]：
2009 – 2011	可以通过符合主要目标的方式完成	无限制
	可以通过符合可选目标的方式完成，并且必须使用 ZEV	0%
	可以通过符合可选目标的方式完成，并且可以使用 AT PZEV 或 PZEV	无限制
2012 – 2017	必须用 ZEV 完成	0%
	可以使用 TZEV、AT PZEV 或 PZEV 完成	无限制

1. 如果适用，本表中的目标为 1962.1（d）(5)（E）3 条款中规定的要求。

该限制适用于同一车企获取的，或由其它车企获取、该车企最终得到的

NEV 积分。

（B）2009 - 2011 年款大型车企结转的规定。2009 - 2011 年款间大型车企 ZEV（不包括 NEV）产生的多余积分，包括从另一方购买的积分，可在随后两个年款中，结转履行 1962.1（b）(2)（B）1. b. 和 （b）(2)（D）条款中规定的 ZEV 最低目标。从第三个年款开始，对于上述方式获得的积分，车企将不能用于履行只能用 ZEV 积分才能满足的 ZEV 目标比例；但是，上述积分仍可以用来履行车企其他可用 TZEV、AT PZEV 或 PZEV 积分来履行的 ZEV 目标比例。例如，2010 年款获取的 ZEV 积分将保留其完整有效性至 2012 年款，此后，该积分只能作为 TZEV、AT PZEV 或 PZEV 积分进行使用。

（C）2009 - 2011 年款除大型车企之外的车企积分结转的规定。2009 - 2011 年款，非大型车企 ZEV（不包括 NEV）产生的多余积分，可由该车企结转，直至该车企在 1962.1（b）(7)（A）条款允许的过渡期后符合大型车企的目标；当车企应当履行大型车企的目标时，该车企必须遵守 1962.1（g）(6)（B）条款中的规定。

对于非大型车企向其它包括大型车企在内的车企转售积分，应当满足 1962.1（g）(6)（B）条款中的规定，积分只能从该积分产生的年款开始转售（例如，公历 2010 年交易的 2009 年款积分只能在 2011 年款结束之前用来履行车企必须用 ZEV 积分履行的目标；从 2012 年款开始，上述积分只能用来满足车企可用 TZEV、AT PZEV、或 PZEV 履行的 ZEV 目标）。

（D）I. 5x 和 IIx 型 ZEV。交付出售或投入运行的 I. 5x 和 IIx 型车辆获取的积分最多可用来抵扣车企应使用 ZEV 积分履行目标的 50%。

（7）补足 ZEV 赤字的要求。

（A）概述。如果车企生产制造并在加州交付销售的 ZEV 少于目标数量，则应在第三个年款结束前，向执行官员提交 2009 - 2014 年款 ZEV 产生的以克每英里为单位计量的积分，以及 2015 - 2017 年款 ZEV 产生的积分来补足上文中所欠 ZEV 积分。提交的积分总量通过以下方式计算：[i] 将车企在某年款内生产并在加州交付销售的 ZEV 积分与车企在某年款内生产并在加州交付销售的 PZEV 的配额数量相加（对于大型车企而言，该积分不能超过 1962.1（b）(2) 条款中规定的数值），[ii] 车企在某年款需要达到的 ZEV 积分目标减去上文中的总量，[iii] 在该车企出现积分数量不足的年款间，针对 PC 和 LDT1 的平均目标和上述所得的值相乘。对于车企通过交付销售 I. 5x 和 IIx 型 ZEV、TZEV、NEV、AT PZEV 和 PZEV 获取的积分，不能被用来补足该车企的 ZEV 赤字；只有 ZEV 产生的积分才能补足该车企所欠 ZEV 积分。

（8）未能满足 ZEV 目标的惩罚措施。如果车企未能生产制造并在加州交付销售所需的 ZEV 数量，且在 2009 – 2014 年款和在 2015 – 2017 年款间未提交相应的积分，同时，该车企亦未在 1962.1（g）(7)(A) 条款中规定的时间段内补足 ZEV 赤字，那么，车企将面临《健康与安全法》第 43211 节民事罚款的处罚，该法案主要针对那些销售车辆但车辆不符合加州管理当局制定的相关排放标准的车企。如果车企在 1962.1（g）(7)(A) 条款规定的最后时限仍未补足所欠 ZEV 积分，则可以认定该车企已经触犯了上述规定，可对其进行处罚。

根据《健康和安全法》第 43211 节，不符合加州管理当局制定标准的车辆数量应等于车企的积分赤字，如果车企应当履行的 ZEV 目标（可以用 PZEV 配额和 PZEV 产生的积分履行目标）的比例未超过 1962.1（b）(2) 条款中规定的比例，那么在 2009 – 2014 年款中，上述车辆数量应四舍五入到千分之一位，在 2015 – 2017 年款中，上述车辆数量应四舍五入到百分之一位。上述车辆数量根据下列公式进行计算：

2009 – 2014 年款：

（某一年款需要的积分数量）–（提交的用于履行该年款目标的积分总量）

/（该年款中 PC 和 LDT1 平均目标）

2015 – 2017 年款：

（某一年款需要的积分数量）–（提交的用于履行该年款目标的积分总量）

（h）试验程序。

（1）确定合规性。2008 年 12 月 17 日通过的并于 2012 年 12 月 6 日进行修订的《乘用车、轻型卡车和中型车辆 2009 – 2017 年款零排放车、混合动力电动车的加州尾气排放标准和测试程序》规定了车辆的认证要求和试验程序，特此将上述文件列入本规则内作参照。

（2）NEV 合规性。1962.1（d）(5)(F) 1 下确定合规性的测试程序在 ETA-NTP002（第 3 修订版）《1993 年 5 月 SAE 的实施标准 J1666：电动车加速、爬坡能力及减速试验程序》（2004 年 12 月 1 日）和 ETA-NTP004（第 3 修订版）《电动车等速续驶里程测试》（2008 年 2 月 1 日）中列出，特此将上述两个文件列入本规则内作参考。

（i）ZEV 具体定义。下列定义适用于第 1962.1 章。

（1）"先进技术部分配额零排放车"或"AT PZEV"指在应用 PZEV 早期引入过渡型乘数前，任何配额大于 0.2 积分的 PZEV。

（2）"辅助动力单元"或"APU"指符合 1962.1（c）(2) 条款中的规定，在零排放里程已经完全耗尽的情况下，为 I.5x 或 IIx 型车辆提供电能或机械能的设备。

燃料加热器不符合此处对辅助动力单元的定义。

（3）"电动车"指任何完全使用电池或电池组提供运动所需能量的车辆，或指那些主要利用电池或电池组但仍然使用能够储存由电动机产生能量的飞轮或电容器的，或回馈制动产生的能量的车辆。

（4）"电量耗尽实际里程"或"R_{cda}"指混合动力电动车当零排放能量储存装置耗尽了存储的电能和通过回馈制动产生的电能时，在城市行驶工况下运行的里程。

（5）"常规取整法"指如果下一位的数字为5或更大的数字，则将要保留的前一位数字增大一个量。如果下一位的数字为4或更小的数字，则保留前一位数字不变。

（6）"东部地区联营体"是指密西西比河以东的第177章所列各州。

（7）"电驱动系统"指在车辆正常运行情况下，向驱动轮提供加速转矩的电机以及与此相关联的电子器件。上述电机以及相关联的电子器件不包括下列组件，即在车辆特定使用状态中，可以用作电机，但是其配置只能充当发电机或发动机启动器的组件。

（8）"增强型先进技术部分配额零排放车"指2009-2011年款中每辆积分配额为1.0或更高值，同时使用ZEV燃料但无乘数的PZEV。增强型AT PZEV指过渡型零排放车。

（9）"低速电动车"或"NEV"指那些符合《车辆法》第385.5节或49 CFR 571.500（2000年7月1日订立）中关于低速车辆的定义、同时符合零排放车标准的机动车。

（10）"投入使用"的车辆指已出售或出租给最终用户，而不是出售或出租给交易商或其他分销链商的车辆；同时该车辆已在加州机动车管理局进行注册以在路上行驶。

（11）"比例值"指某车企在加州的适用销量与该车企在第177章所列各州内适用销售量的比值。在任何特定的年款中，必须使用与计算销售量相同的方法对比例值进行计算。

（12）"里程延长式电动车"指那些主要通过零排放能量存储装置提供能量的车辆，纯电续驶里程达75英里以上，除此之外，该车辆还配备备用的辅助动力单元（直到能量存储装置被完全耗尽前，该辅助动力单元不工作），并且符合1962.1（d）（5）（G）条款中规定的要求。

（13）"回馈制动"指将部分损耗于摩擦制动的能量以电流的形式返回到贮能装置中。

（14）"第177章所列各州"指根据联邦《清洁空气法案》第177章的规定（42

USC §7507①) 执行加州 ZEV 目标的美国各州。

（15） "过渡型零排放车"指那些配额积分为 1.0 或更高，同时使用 ZEV 燃料的 PZEV。

（16） "0，I，I.5，II，III，IV 和 V 型 ZEV"定义见 1962.1（d）（5）（A）条款。

（17） "西部地区联营体"是指密西西比河以西的第 177 章所列各州。

（18） "ZEV 燃料"指为公路上零排放车提供牵引能量的燃料。当前技术 ZEV 燃料包括电能、氢能和压缩空气。

（j）缩写。在 1962.1 中使用下述缩略语：

"AER"指纯电续驶里程。

"APU"是指辅助动力单元。

"AT PZEV"指先进技术部分零排放配额车辆。

"CFR"是指联邦法规（美国）。

"DMV"指加州机动车管理局。

"EAER"是指等效纯电续驶里程。

"EAER_{u40}"指具备 40 英里 R_{cda}（城市电量耗尽实际里程）的插电式混合动力电动车能够达到的城市等效纯电续驶里程。

"FR"联邦公报。

"HEV"是指混合动力车。

"LDT"指轻型卡车。

"LDT1"指满载重量为 0 – 3750 磅的轻型卡车。

"LDT2"指满足 LEV II 排放标准的满载重量为 3 751 – 8 500 磅的轻型卡车，或者满足 LEV I 排放标准的满载重量为 3 751 – 5 750 磅的轻型卡车。

"LVM"指大型车企。

"MDV"指中型车辆。

"非甲烷有机气体"或"NMOG"是指被氧化和未被氧化的碳氢化合物的总质量

"NEV"指低速电动车。

"NO_x"指氮氧化物。

"PC"是指乘用车。

"PZEV"指部分配额零排放车，这些车辆在加州交付销售，并有资格获得至少 0.2 分的配额积分。

---

① 译者注：《美国法典》第 42 部分为《清洁空气法案》，第 7507 章为空气质量列于全国空气质量标准地区的新产机动车尾气排放标准。

"R_{cda}"指在城市电量耗尽实际里程。

"SAE"是指（美国）汽车工程师协会。

"SULEV"指特超低排放车。

"TZEV"指过渡型零排放车。

"I.5x 型"指所有纯电续驶里程为 75 到 100 英里的电动车。

"IIx 型"指所有纯电续驶里程为 100 英里或更长的电动车。

"UDDS"指市区测功机驾驶循环。

"UF"是指效用因子。

"US06"指 US06 补充联邦测试程序。

"VMT"是指行驶里程。

"ZEV"指零排放车。

（k）可分割性。本节中的每条规定都是可分割的，如本节的某一条规定被认定为无效，其余的所有规定仍然继续完全有效。

（l）公开披露。委员会拥有符合第 1962.1 节要求的车辆记录，这些记录应当作为公共记录进行如下披露：

（1）2009 - 2017 年款间，每家车企的年产量以及每辆 ZEV（包括 ZEV 各类型）、TZEV、AT PZEV 和 PZEV 产生的积分。

（2）每家车企 2010 - 2017 年款的积分余额。

　　（A）每种车辆的类型：ZEV（不包括 NEV）、I.5x 和 IIx 型车辆、NEV、TZEV、AT PZEV 和 PZEV；

　　（B）先进技术示范项目；

　　（C）交通系统；以及

　　（D）根据 1962.1（d）（5）（C）条款获取的积分，包括从第三方处获取或转移的积分。

　　注意：引用的规定章节：《健康和安全法》第 39600、39601、43013、43018、43101、43104 和 43105 章。参考：《健康和安全法》第 38562、39002、39003、39667、43000、43009.5、43013、43018、43018.5、43100、43101、43101.5、43102、43104、43105、43106、43204、43205、43205.5 和 43206 节。

§1962. 2 2018 及后续年款乘用车、轻型卡车和中型车辆的零排放车标准

（**a**）**零排放车（ZEV）排放标准**。如果车辆在任何可能的行驶状态或条件下排放（不包括空调系统的排放）的任何标准污染物（或前体物）为零，执行官员将确认 2018 及后续年款新产乘用车（PC）、轻型卡车（LDT）和中型车辆（MDV）符合零排放车标准。

（**b**）**ZEV 比例目标**。

（**1**）**ZEV 积分基本比例目标**。

（**A**）**基本目标**。根据 1962. 2（b）条款，下表中列出了每家车企需达到的最低零排放车比例，此比例用车企生产制造并在加州交付销售的乘用车（PC）和轻型卡车（LDT）的零排放车计算。ZEV 目标将依据相应年款的非甲烷有机气体（NMOG）排放报告确定。

年款	积分比例目标
2018	4. 5%
2019	7. 0%
2020	9. 5%
2021	12. 0%
2022	14. 5%
2023	17. 0%
2024	19. 5%
2025 及后续年款	22. 0%

（**B**）**计算 ZEV 数量**。针对 2018 及后续年款，某年款的产量应根据该车企之前第二、三、四年生产并在加州交付销售 PC 和 LDT 的平均产量来计算（例如 2019 年款的 ZEV 目标是根据 2015 - 2017 年款 PC 和 LDT 的平均产量为基础计算的）。年款平均产量只用于确定 ZEV 目标，不用于确定车企规模（比如三年平均法）。在确定 ZEV 目标的过程中，某车企（例如车企 A）生产了一辆 PC 或 LDT，但是由另外一家车企（例如车企 B）以其自身品牌在加州推广销售，则该车应算在车企 B 名下。

1.［保留］。

2.［保留］。

3. 车企可以向执行官员申请采用其生产并在加州交付销售的某年款 PC 及 LDT 数量来计算 ZEV 目标（当年预测法），代替上文提及的三年平均法，当年预测法最多可连续使用 2018－2025 年款中的两个年款。为使用同年款计算法，车企在给执行官员的申请中，必须表明该车企生产并在加州交付销售的 PC 和 LDT 数量由于不可预见和无法控制的情况较去年至少减少 30%。

（C）［保留］。

（D）确定车企销售量时排除 ZEV。使用 1962.2（b）（1）（B）条款中任一方法计算车企适用销售量时，车企应排除其生产并在加州交付销售的，以及该车企控股 33.4% 以上的附属公司生产的 NEV（低速电动车）数量。

（2）大型车企的目标。

（A）［保留］。

（B）［保留］。

（C）［保留］。

（D）［保留］。

（E）2018－2025 年款大型车企的目标。大型车企必须产生与最低 ZEV 比例目标相当的 ZEV 积分，如下所示。车企可以使用 TZEV（过渡型零排放车）获得的积分履行剩下的 ZEV 目标，列举如下。

年款	总的零排放车比例目标	最低零排放车比例	过渡型零排放汽车比例
2018	4.5%	2.0%	2.5%
2019	7.0%	4.0%	3.0%
2020	9.5%	6.0%	3.5%
2021	12.0%	8.0%	4.0%
2022	14.5%	10.0%	4.5%
2023	17.0%	12.0%	5.0%
2024	19.5%	14.0%	5.5%
2025	22.0%	16.0%	6.0%

（F）2026 及后续年款大型车企的目标。针对 2026 及后续年款，车企必须履行 22% 的 ZEV 积分比例目标。在车企履行积分比例目标时，使用 TZEV 积分的上限为在加州 PC 和 LDT 适用产量的 6%。剩余比例目标必须使用 ZEV 积分履行。

（3）中型车企的目标。对于 2018 及后续年款，根据 1962.2（b）条款，中型车企可以使用从 TZEV 处获取的积分来履行车企所有 ZEV 积分比例目标。

（4）小型车企的目标。小型车企不必履行 ZEV 积分比例目标。但车企可以获得、存储、交易和买卖其生产并在加州交付销售的 ZEV 和 TZEV 获得的积分。

（5）〔保留〕。

（6）〔保留〕。

（7）小型车企和中型车企的规模变化。

（A）销售于加州产量增加的情况。针对 2018 及后续年款，如果车企先前连续三个三年生产并在加州交付销售的新产 PC、LDT 以及 MDV 平均产量超出 4 500 辆（即三年期总产量超过 13 500 辆），该车企应不再被视为小型车企，该车企必须从最近上一个三年后的第一个年款开始履行中型车企的 ZEV 目标。例如，如果（小型车企）车企 A 在其 2018 - 2020、2019 - 2021 以及 2020 - 2022 年款中，生产的 PC、LDT 和 MDV 车辆的平均数都超过了 4 500 辆，则该车企从 2023 年款开始，应履行中型车企的目标。

如果车企先前连续三个三年生产并在加州交付销售的新产 PC、LDT 以及 MDV 平均产量超出 20 000 辆（即总产量在三年期间超过 60 000 台），车企应不再被视为中型车企，同时必须从最近上一个三年后的第一个年款开始履行大型车企的 ZEV 目标。例如，如果（中型车企）车企 B 在其 2018 - 2020、2019 - 2021 以及 2020 - 2022 年款中，生产的 PC、LDT 和 MDV 车辆的平均数都超过了 20 000 辆，则该车企从 2023 年款开始应履行大型车企的目标。

当某车企在 2018 或以后年款中由于股权变化而不再是小型或中型车企时，该车企在履行本条款规定的新要求时，都是从最近上一个三年后的第一个年款开始执行。

（B）销售于加州产量减少的情况。如果车企先前连续三个三年生产并在加州交付销售的 PC、LDT1 和 LDT2，以及 MDV 平均产量低于 4 500 辆或 20 000 辆，则该车企应被视为小型或中型车企，同时该车企必须履行针对小型或中型车企的 ZEV 目标，从最近上一个三年后的第一个年款开始执行。例如，如果某车企 C 在其 2019 - 2021、2020 - 2022 以及 2021 - 2023 年款中 PC、LDT 以及 MDV 的平均产量少于 20 000 辆，则该车企从 2024 年款开始应履行中型车企的目标。

（C）变更所有权时加州产量的计算。车企在某年款发生任何所有权变更，将会影响车企下一年款需履行的目标。如果车企在所有权变更时同时生产两种年款车型，下一年款的规模应依据较早年款来决定。小型车企或中型车企下一年款的规模应依据其先前连续三年在加州的平均产量判断。例如，某一车企在 2019 年发生所有权变化，该车企当年同时生产 2019 和 2020 年款车辆，此时车企 A 的产量应当与车企 B 的产量合计，在上述情况中，车企 A 2020 年款的规模将由车企 A 和 B 2017 -

2019 年款产量来决定。如果车企 A 2019 年款的产量必须与车企 B、C 的产量合计，在此年款中，由于发生的股权变化导致车企 B 与车企 A 合并产量的要求被取消，那么对于 2020 年款规模，车企 A 将以 2017 - 2019 年款中车企 A 和 C 的产量为基础确定。在任一情况下，1962.2（b）(7)(A) 及（B）条款中的规定将会适用。

（c）过渡型零排放车（TZEV）。

（1）说明。本部分 1962.2（c）设定了在加州交付销售的车辆是否属于 TZEV 的标准。

（2）TZEV 目标。为了使车辆符合获取 TZEV 配额的标准，车企必须证明车辆符合以下所有要求：

（A）SULEV（特超低排放车）标准。证明车辆符合 1961.2（a）(1) 条款规定的针对 PC 和 LDT 的 15 万英里 SULEV 20 或 30 尾气排气标准。双燃料，灵活燃料及混合燃料车辆必须证明在使用两种燃料的情况下符合 15 万英里 SULEV20 或 30 尾气排气标准。车企需证明 2018 - 2019 年款的 TZEV 符合 1961（a）(1) 中针对 PC 和 LDT 的 15 万英里 SULEV 尾气排放标准。

（B）挥发性排放。证明车辆符合 1976（b）(1)(G) 或 1976（b）(1)(E) 条款中的挥发性排放标准；

（C）车载诊断系统（OBD）。如果适用，证明车辆满足 1968.1 或 1968.2 条款中的 15 万英里车载诊断要求；并且

（D）延长的质保期。延长 2037（b）(2) 和 2038（b）(2) 条款中规定的性能和缺陷保修期至 15 年或 15 万英里（以先达到者为准），对于使用零排放能源存储设备（如电池、超级电容或其他电能存储设备）作牵引动力的车辆，保修为 10 年期。

（3）TZEV 配额

（A）TZEV 零排放行驶里程配额计算。满足 1962.2（c）(2) 条款的规定并且具有零排放行驶里程（VMT）的车辆，其行驶里程根据 2012 年 3 月 22 日通过的并于 2012 年 12 月 6 日进行修订的《乘用车、轻型卡车和中型车辆 2018 及后续年款零排放车、混合动力电动车的加州尾气排放标准和测试程序》进行定义和计算，并作为等效纯电续驶里程（EAER）进行测量，并根据下面的公式计算配额：

UDDS 测试周期里程（纯电续驶里程）	配额
<10 英里纯电动里程	0.00
≥10 英里纯电动里程	TZEV 积分 = ［（0.01）× EAER + 0.30］
>80 英里（积分上限）	1.10

1. US06 能力配额。具备至少 10 英里 US06 纯电续驶里程的 TZEV 将获得额外 0.2 分的配额。US06 测试周期里程能力应根据《乘用车、轻型卡车和中型车辆 2018 及后续年款零排放车、混合动力电动车的加州尾气排放标准和测试程序》（2012 年 3 月 22 日通过，上次修订时间为 2012 年 12 月 6 日）第 G.7.5 节确定。

（B）〔保留〕。

（C）〔保留〕。

（D）〔保留〕。

（E）氢内燃机汽车积分。一辆满足 1962.2（c）(2) 的规定，同时具备在 UDDS 工况下至少 250 英里里程的氢内燃机车辆将获得 0.75 分的配额，此处获取的配额可以和根据 1962.2（c）(3)（A） 获取的配额一起获取，总积分不能超过 1.25 分。

（d）ZEV 积分条件。

（1）〔保留〕。

（2）〔保留〕。

（3）〔保留〕。

（4）〔保留〕。

（5）2018 及后续年款的 ZEV 积分。

（A）ZEV 积分计算。交付销售的零排放车获取的积分是以零排放车在 UDDS 工况下的行使里程为基础使用以下等式计算的，行使里程数是根据《乘用车、轻型卡车和中型车辆 2018 及后续年款零排放车、混合动力电动车的加州尾气排放标准和测试程序》（2012 年 3 月 22 日通过，上次修订时间为 2012 年 12 月 6 日）进行计算的：

$$ZEV\ 积分 = （0.01）× （UDDS\ 里程）+0.50$$

1. 在 UDDS 工况下少于 50 英里里程的 ZEV 将获得零积分。

2. 每辆零排放车根据 1962.2（d）(5)（A） 获取的积分最多为 4 分。

（B）〔保留〕。

（C）〔保留〕。

（D）〔保留〕。

（E）计算在加州和第 177 章所列各州中投入使用的特定 ZEV 数量。利用氢燃料电池车（这些车辆满足加州零排放车标准，并在加州和第 177 章所列各州中交付销售并投入使用）获得积分的大型和中型车企，被认为在加州和第 177 章所列各州履行了 1962.2（b）条款中规定的 ZEV 比例目标。车企获取的积分与某比率相乘，该比率是车企根据 1962.2（b）(1)（B）条款的规定，在获取积分所在州的适用产量与在加州适用产量的比值（以下称比例值）。车企在第 177 章所列各州中获得的积分将以比例值获取，在加州以全值获取。

1. 第 177 章所列各州的可选达标途径。

a. 减少的 ZEV 和 TZEV 百分比。那些根据 1962.1（d）(5)（E）3 条款规定充分遵守第 177 章所列各州的可选达标途径的大型和中型车企，可在第 177 章所列各州承担减少的 ZEV 比例和可选 TZEV 比例目标（根据 1962.2（b）(2)（E）条款从最低 ZEV 比例和 TZEV 比例中减少），此比例等于下述根据 1962.2（b）(1)（B）条款计算的销售量的百分比。

ZEV

年款	2018	2019	2020	2021
现有的最低 ZEV 比例	2.00%	4.00%	6.00%	8.00%
第 177 章所列各州可选达标途径的调整	62.5%	75%	87.5%	100%
第 177 章所列各州 ZEV 最低目标	1.25%	3.00%	5.25%	8.00%

TZEV

年款	2018	2019	2020	2021
现有的 TZEV 比例	2.50%	3.00%	3.50%	4.00%
第 177 章所列各州可选达标途径的调整	90.00%	100%	100%	100%
第 177 章所列各州的新 TZEV 比例	2.25%	3.00%	3.50%	4.00%

总比例目标

年款	2018	2019	2020	2021
第 177 章所列各州可选新总比例目标	3.50%	6.00%	8.75%	12.00%

i. 在西部地区联营体和东部地区联营体内交易和转移 ZEV 和 TZEV 积分。那些根据 1962.1（d）(5)（E）3 条款遵守第 177 章所列各州的可选达标途径目标的车企，可以在西部地区联营体内交易和转移某年款的 ZEV 和 TZEV 积分以履行 1962.2（d）(5)（E）2.a 条款中规定的目标，并且积分不会产生溢价。例如，某车企为弥补 2019 年款在 X 州的 100 个积分差额，该车企可以从西部地区联营体内的 Y 州转移出 100 个（2019 年款）ZEV 积分。那些根据 1962.1（d）(5)（E）3 条款遵守第 177 章所列各州的可选达标途径的车企可以在东部地区联营体内交易和转移特定年款的 ZEV 和 TZEV 积分，来满足 1962.2（d）(5)（E）2.a 中规定的要求，并且积分不会产生溢价。例如，某车企为弥补 2019 年款在 W 州的 100 个积分差额，该车

企可以从东部地区联营体内的 Z 州转移出 100 个（2019 年款）ZEV 积分。

ii. 在西部地区联营体和东部地区联营体之间交易和转移 ZEV 和 TZEV 积分。那些根据 1962.1（d）(5)（E）3 条款遵守第 177 章所列各州的可选达标途径目标的车企，可以在西部地区联营体和东部地区联营体之间交易和转移某年款 ZEV 和 TZEV 积分以履行 1962.2（d）(5)（E）2.a 条款中规定的目标，但转移积分将产生其本身价值 30% 的溢价。例如，某车企为弥补 2019 年款在西部地区联营体中的 100 个积分差额，该车企可以从东部地区联营体中转移出 130 个（2019 年款）积分。车企不得从加州 ZEV 银行向东、西部地区联营体中交易或转移积分，或者是从东、西部地区联营体向加州 ZEV 银行中交易或转移积分。

b. 报告要求。以年为基础，截至年款结束后的公历年 5 月 1 日，根据 1962.1（d）(5)（E）3 规定选择第 177 章中所列各州可选达标途径的车企应当以书面形式向执行官员，以及第 177 章所列各州提交报告，包括详细项目单，表明车辆在东、西部地区联营体中交付使用的所在地。详细项目单应包括以下内容：

i. 根据 1962.2（b）(1)（B）条款，在地区联营体内第 177 章所列各州交付销售的 PC 和 LDT 的总适用产量。

ii. 构造、型号、车辆识别码、获取的积分以及车企为履行 1962.2（d）(5)（E）2.a. 条款 ZEV 目标而在第 177 章所列各州交付销售并投入使用 TZEV 和 ZEV 的州名。

c. 未能履行第 177 章所列州的可选达标途径的目标。根据 1962.1（d）(5)（E）3 条款选择第 177 章所列各州可选达标途径的车企，如果某年款未能履行 1962.2（d）(5)（E）2.a. 条款中调整的比例目标或未能使用 1962.2（g）(7)（A）条款允许的特定积分在特定时间内补足赤字，则在 2018－2021 年款间，在第 177 章所列各州中，应当履行 1962.2（b）中总的 ZEV 比例目标，并且 1962.2（d）(5)（E）2.a. 中的联营体规定不适用。如果车企未能遵守，任何在第 177 章所列各州之间的 ZEV 或 TZEV 积分转移都是无效的，ZEV 或 TZEV 积分将返回积分所在州。如果车企未能在 2018 年款结束之前补足 ZEV 赤字，其处罚应根据其所处的第 177 章所列的州单独计算。

d. 1962.2 条款中的规定，除了 1962.2（d）(5)（E）2 中特殊修改过的，均适用于选择第 177 章所列各州的可选达标途径的车企。

（F）NEV。低速电动车必须满足以下条件后，才能获得 0.15 积分：

1. 技术参数。当低速电动车符合下列所有条件时，NEV 可以获得积分：

a. 加速度。在至少载有 332 磅有效载荷、电池在 50% 荷电状态下启动时，车辆在 6 秒或更短时间内完成 0－20 英里每小时加速。

b. 最高速度。在至少载有 332 磅有效载荷、电池在 50% 荷电状态下启动时，车辆的最高速度下限为 20 英里每小时。按照 49 CFR 571.500（美国联邦法规）(68 FR 43972，2003 年 7 月 25 日）进行测试时，车辆最高速度不应超过 25 英里每小时。

c. 等速续驶里程。在至少载有 332 磅有效载荷、启动时电池处于 100% 荷电状态下，并以最高速度等速运行时，最低续驶里程为 25 英里。

2. 电池要求。低速电动车必须配备一个或多个密封、免维护电池。

3. 保修要求。NEV 的动力系统，包括电池组在内，保修期必须至少为 24 个月。NEV 保修期的前六个月内必须完全保修，剩余的保修期是可选延长保修（可购买）以及可按比例进行保修。如果按比例延长保修，保修或退还的电池组初始价值的百分比至少不低于剩余的按比例分配的保修期比例。为了便于进行此类计算，电池组的寿命必须以时间间隔来表示，单个时间间隔不大于三个月。或者，车企可以在整个延长的保修期限内负责电池组 50% 初始价值的保修服务。

批准积分之前，执行官员可以要求车企提供具有代表性的车辆和电池保修单的副本。

4. NEV 充电要求。低速电动车必须满足 1962.3（c）(2）条款中规定的充电要求。

（G）里程延长式电动车（BEV$_X$）。根据 1962.2（d）(5)（A）条款规定，基于 UDDS 工况下的纯电续驶里程，BEV$_X$ 必须满足以下条件才能获得积分：

1. 排放要求。电动车必须满足 1962.2（c）(2)（A）至（D）条款中的所有 TZEV 目标。

2. 辅助动力单元的操作。辅助动力单元首次启动并进入"电量维持混合运行"后，UDDS 工况下的续驶里程必须小于或等于汽车辅助动力单元启动前的 UDDS 工况下的纯电续驶里程。车辆的辅助动力单元不能在用户可选驱动模式下启动，除非用于牵引的能量储存系统完全耗尽。

3. 最小零排放里程要求。BEV$_X$ 必须具备至少 75 英里的 UDDS 工况下的纯电续驶里程。

（e）［*保留*］。

（f）［*保留*］。

（g）积分的产生和使用；处罚的计算。

(1) 说明。如果车企生产并在加州交付销售的某年款 ZEV 和 TZEV 数量超过 1962.2（b）规定的 ZEV 目标，该车企将根据 1962.2（g）的规定获得积分。

(2) ZEV 积分计算。

(A) ZEV 积分。车企在某年款获得的 ZEV 积分总量应以积分为单位进行表示。该积分总量等于车企生产并在加州交付销售的 ZEV 积分总量，减去车企生产并在加州交付销售用于满足目标的 ZEV 积分；或者 1962.2（d）(5)(E) 2.a. 条款中规定的积分目标后所得的数值。

(B) TZEV 积分。车企在某年款获得的 TZEV 积分总量应以积分为单位进行表示，该积分总量等于车企生产并在加州交付销售的 TZEV 获得的 ZEV 配额总量，减去车企生产并在加州交付销售用于满足 ZEV 目标的 TZEV 积分；或者 1962.2（d）(5)(E) 2.a. 条款中规定的目标。

(C) 独立的积分账户。车企从 ZEV、BEVx、TZEV 和 NEV 中获取的积分应分别保管在不同的帐户中。

(D) 舍入的积分。ZEV 积分和借记须使用传统舍入法将最后的积分和借记总数舍入到最近 1/100 位。

(3) 针对 MDV 的 ZEV 积分。被列为 MDV 的 ZEV 和 TZEV 获取的积分可以满足针对 PC 和 LDT 的 ZEV 目标，并根据 1962.2（g）的规定包含在 ZEV 积分的计算之中。

(4) 先进技术示范项目的 ZEV 积分。

(A) ［保留］。

(B) ZEV。被列为小型或中型车企在加州先进技术示范项目中两年或两年以上的 ZEV，包括 BEVx，但不包括 NEV，即使该车辆没有"交付销售"或向加州机动车管理局登记过，仍可获得 ZEV 积分。要取得这样积分，车企必须合理满足执行官员的要求，即证明车辆会经常使用，可以评估与安全事宜、基础设施、燃料规格或公众教育相关方面的问题，并且在最初两年 50% 及以上的时间将在加州运行。上述情况中的车辆有资格获得与该车辆如果交付销售，或者燃料电池车投入使用的情况下获得的积分相同的积分。根据 1962.2（g)(4) 条款，车企每年在每个第 177 章所列各州每个车型最多可有 25 辆车获取积分。超过 25 辆上限的车辆将没有资格获得该类型积分。

(5) 交通系统的 ZEV 积分。

(e) ［保留］。

(f) ［保留］。

(g) 交通系统积分使用上限。

1. ZEV 。根据 1962.1（g）（5）条款的规定，通过 ZEV 或 BEVx 获得或分配的交通系统的积分，不包括车辆本身获得的任何积分，最多可用来满足车企指定年款 ZEV 目标的十分之一，以及最多可用来满足必须用 ZEV 履行的 ZEV 目标的十分之一（见 1962.2（b）（2）（E）条款的规定），或者 1962.2（d）（5）（E）2.a. 条款中的目标。

2. TZEV 。根据 1962.1（g）（5）的规定，通过 TZEV 获得或分配的交通系统的积分，不包括车辆本身获得的任何积分，最多可用来满足车企可以使用 TZEV 履行的 ZEV 目标部分的十分之一，或者，如果适用的话，可用 TZEV 来履行 1962.2（d）（5）（E）2.a. 规定的车企其他目标部分的十分之一。但是上述获得的积分只能以和该类别车辆获得的其他积分相同的使用方式进行使用。

（6）ZEV 积分的使用。车企可以通过向执行官员提交与 1962.2（b）规定中等量的 ZEV 积分来满足某一特定年款的 ZEV 目标。每个类别的积分可以用来满足针对该类别的目标，也可以用来满足比该类别获取的 ZEV 积分更少的那些类别的目标，但是不能用于满足比该类别获取的 ZEV 积分更多的那些类别的目标，但 PZEV 和 AT PZEV 折扣积分除外。例如，TZEV 积分可用于满足 TZEV 积分目标，但不能用于满足只能用 ZEV 的积分才能满足的部分。这些积分可以由车企预先获取或从另一方处获得。

（A）PZEV、AT PZEV 的折扣积分和 NEV 积分的使用。2018 – 2025 年款，PZEV、AT PZEV 的折扣积分和 NEV 积分最多可抵扣能用 TZEV 积分履行的车企目标部分的（按照 1962.2（d）（5）（E）2.a. 规定可通过 TZEV 积分来满足车企目标部分的）四分之一。2018 – 2019 年款，中型车企可利用 PZEV、AT PZEV 的折扣积分和 NEV 积分来满足其全部目标。这些积分可以由车企预先获取或从另一方处获得。2025 年款后，PZEV、AT PZEV 的折扣积分不能再使用。

（B）BEV$_x$积分的使用。BEV$_x$积分最多可用来满足必须用 ZEV 积分履行的车企目标的 50%。

（C）温室气体（GHG）– ZEV 履约积分。

1. 应用。根据下面的条件，在 2016 年 12 月 31 号以前，车企可以向执行官员申请获得本条款规定的积分：

a. 根据 1961.1 和 1961.3，车企不能有 2017 年款的履约借记以及之前所有年款的未履约的借记。或者，根据 1961.1（a）（1）（A）（ii）和 1961.3（c），必须证明符合国家温室气体排放计划的标准。

b. 根据 1962.1，车企不能有 2017 年款的履约借记以及之前所有年份的未履约的借记。

c. 车企必须提交预定产品计划的文件，以显示其超额完成 1961.3 规定的目标，或者超额完成 1961.3（c）中规定的国家温室气体排放计划的履约目标，即从 2018－2021 年款，每年 CO_2 的排放量超额达到至少 2.0 克 CO_2/英里的标准，并且承诺每年都会这样达标。

2. 生成和计算积分。基于以往年款履约情况，针对 2018－2021 年款，车企可计算其超额完成 1961.3 规定的目标，或是超额完成 1961.3（c）规定的国家温室气体排放计划的目标。例如，为了计算 2018 年款 1962.2（g）(6)(C) 条款规定的超额完成的积分，车企会基于 2017 年款超额履约 1961.3 条款的目标情况，以及超额履约 1961.3（c）规定的国家温室气体排放计划的目标。

a. 至少每年超额履行 1961.3 条款的目标或 1961.3（c）规定的国家温室气体排放计划目标达到 2.0 克 CO_2/英里排放量标准。下面的方程是用来计算获得的 ZEV 积分，以便符合 1962.2（g）(6)(C) 的规定：

［(车企在美国 PC 和 LDT 销售量）×（克 CO_2/英里给定年款低于车企 GHG 标准的部分）］/(给定年款车企 GHG 标准)

b. 根据 1961.3（a)(9) 获得的积分，或者根据 40 CFR，86 部分，S 分部分的 § 86.1866－12（a），§ 86.1866－12（b）和 § 86.1870－12 中获得的积分可以不包括在上述计算克 CO_2/英里积分的等式中。上述计算中涉及的 ZEV 必须包含 1961.3 规定的上游排放值。

c. 根据 1961.1 和 1961.3 结转的以往年款的克 CO_2/英里积分或者根据 1961.3（c）超额完成国家温室气体排放计划结转的积分，可以不包括在上述计算克 CO_2/英里积分的等式中。

3. GHG-ZEV 履约积分的使用。使用根据 1962.2（g)(6)(C) 获得的积分，车企可以使用不超过下列表格中列举的比例来满足全部 ZEV 目标或者必须用 ZEV 积分履行的 ZEV 目标的部分。

2018	*2019*	*2020*	*2021*
50%	50%	40%	30%

在某年款中，按照 1962.2（g)(6)(C) 获得的积分只能用于适用的年款中，而不能用于其他年款中。

用于计算在本条规定下的 GHG－ZEV 超额履约积分的克 CO_2/英里积分，必须从车企的 GHG 履约银行中扣除，并且不能结转以在将来履行 1961.3 的目标或者 1961.3（c）规定下的国家温室气体排放计划的目标。

4. 报告要求。要求车企每年都要提交符合 1962.2（g）(6)（C）规定的积分计算结果，任何 1961.3 规定下或者 1961.3（c）规定下的国家温室气体排放计划下的以往年款的剩余积分/借记，以及 1961.3 规定下的或 1961.3（c）和本节 1962.2（g）(6)（C）规定下的 2021 年款前的预期积分/借记。

如果车企已经有能力获得 1962.2（g）(6)（C）规定下的积分，却在某一年款中未达到 2.0 克 CO_2/英里的标准，则在当前年款和未来年款中，车企必须要完全满足 ZEV 目标，并且不能在任一年款中根据 1962.2（g）(6)（C）规定获取积分。

（7）补足 ZEV 赤字的要求。

（A）概述。在某年款中，如果车企生产并在加州交付销售的 ZEV 数量比目标少，那么在下一年款之前，应向执行官员提交一定数量的积分来补足赤字。需要提交的积分将按照以下标准计算：［i］把车企生产并在加州交付销售的 ZEV 积分数，和车企生产并在加州交付销售的 TZEV 积分数相加（对于大型车企来说，不要超出 1962.2（b）(2）所规定的数量），［ii］从车企应该生产并在加州交付销售所得积分数中减去上述总和。BEVx、TZEV、NEV 或改装的 AT PZEV 及 PZEV 积分不允许用来弥补车企的 ZEV 赤字；只有从 ZEV 获得的积分才能用于弥补车企的 ZEV 赤字。

（8）未能满足 ZEV 目标的惩罚措施。任何车企，如其生产并在加州交付销售 ZEV 数量未达到目标并且没有提交适当积分数量或没有在 1962.2（g）(7)（A）规定的特定时间内补足 ZEV 赤字，则将面临《健康与安全法》第 43211 章的民事罚款，该法案主要针对那些销售新款车辆但不符合加州管理当局制定的相关排放标准的车企。如果车企在 1962.2（g）(7)（A）条款规定的最后时限前仍未补足所欠 ZEV 积分，则可认定该车企已经触犯了上述规定。根据《健康与安全法》第 43211 章的规定，不符合加州管理当局标准的车辆数量应等于车企的积分赤字，并且四舍五入到最近的百分之一，具体根据下面的公式计算得出。在某年款中，如果可以用 TZEV 或从这些车辆获得的积分来满足车企 ZEV 目标，但不能超过 1962.2（b）(2）中允许的比例。

（某一年款需要的积分数量）－（提交的用于履行该年款目标的积分总量）

（h）试验程序。

（1）确定合规性。确定是否履约 1962.2 需要的认证要求和试验程序可参考《乘用车、轻型卡车和中型车辆 2018 及以后年款零排放车、混合动力电动车的加州尾气排放标准和测试程序》（2012 年 3 月 22 日通过并于 2012 年 12 月 6 日进行最后修订），特此将上述文件列入本规则内作参照。

（2）NEV 合规性。1962.1（d）(5)（F）1 下确定合规性的测试程序可参见

ETA-NTP002（第3修订版）《1993年5月SAE实施标准J1666：电动车加速、爬坡能力及减速试验程序》（2004年12月1日）和ETA-NTP004（第3修订版）《电动车等速续驶里程测试》（2008年2月1日）中列出，特此将上述文件列入本规则内作参照。

（i）ZEV具体定义。下列定义适用于1962.2章节：

（1）"辅助动力单元"或"APU"，指符合1962.2（c）（2）条款中的规定，在零排放量程已经完全耗尽的情况下，能够为BEVx提供电能或机械能的设备。燃料加热器不符合此处对辅助动力单元的定义。

（2）"电量耗尽实际里程"或"R_{cda}"指当零排放能量储存装置耗尽了存储的电能和通过回馈制动产生的电能时，混合电动车在城市行驶工况下运行的里程。

（3）"常规取整法"指如果下一位的数字为5或更大的数字，则将要保留的前一位数字增大一个量。如果下一位的数字为4或更小的数字的，则保留前一位数字不变。

（4）PZEV和AT PZEV折扣积分是指，根据章节1962和1962.1的规定，通过交付销售PZEV和AT PZEV得到的积分，并按照1962.1（g）（2）（F）的规定进行了打折。

（5）"东部地区联营体"是指密西西比河以东的第177章所列各州。

（6）"能源存储设备"是指一种存储设备，可以以最小的能源存储能力提供发动机启停、牵引增强、回馈制动和（名义上的）充电维持模式驱动能力。对于TZEV，相对于经证明的新车里程能力，不需要说明最小的里程范围。

（7）"氢燃料电池车"是指主要燃料是氢，但也有外接充电功能。

（8）"氢燃料内燃机车"是指仅仅用氢气做燃料的TZEV。

（9）"过半数所有权"是指按照CCR①第1900章节的规定，在决定规模大小时，某个车企占据另一个车企33.4%以上的所有权。

（10）"车企在美国PC和LDT销售量"是指车企在某年款中，在美国销售的PC和LDT的总量。

（11）"低速电动车"或"NEV"指那些符合《车辆法》第385.5节或49 CFR 571.500（2000年7月1日成立）中关于低速车辆定义、同时符合零排放车标准的那些机动车。

（12）"投入使用"的车辆指已出售或出租给最终用户，而不是出售或出租给交易商或其他分销链商的车辆；同时该车辆已经在加州机动车管理局注册以在道路上

---

① 译者注：《加州法规》

行驶。

（13）"比例值"指某车企在加州的适用销量和该车企在第177章所列州内适用销量的比值。在某年款中，必须使用与计算销量相同的方法对比例值进行计算。

（14）"里程延长式电动车"指那些主要通过零排放能量存储装置提供能量的车辆，纯电里程为至少75英里，除此之外，车辆还配备备用的辅助动力单元（直到能量存储装置被完全耗尽前，该辅助动力单元不工作），并且符合1962.2（d）（5）（G）条款中规定的要求。

（15）"第177章所列各州"指根据《联邦清洁空气法案》第177章的规定（42 USC § 7507）执行加州ZEV目标的美国各州。

（16）"过渡型零排放车"或"TZEV"是指符合1962.2（c）（2）所有标准，有资格取得1962.2（c）（3）（D）或（E）配额的车辆。

（17）"西部地区联营体"是指密西西比河以西的第177章所列各州。

（18）"零排放车"或"ZEV"是指不排放任何标准污染物（或前体污染物），或是在任何行驶状态或条件下不产生温室气体的车辆。

（19）"ZEV燃料"是指为公路上零排放车提供牵引能量的燃料。当前技术条件下ZEV燃料包括电能、氢能和压缩空气。

（j）缩写。在1962.2章中使用以下缩略语：

"AER"是指纯电续驶里程。

"APU"是指辅助动力单元。

"AT PZEV"是指先进技术部分零排放配额车。

"BEV_x"是指里程延长式电动车。

"CFR"是指联邦法规（美国）。

"CO_2"是指二氧化碳。

"DMV"指加利福尼亚州机动车管理局。

"EAER"是指等效纯电续驶里程。

"FR"是指联邦公报。

"g"是指克。

"HEV"是指混合动力车。

"LDT"指轻型卡车。

"LDT1"指总质量为0-3750磅的轻型卡车。

"LDT2"指满足LEV II排放标准的总质量为3 751-8 500磅的轻型卡车，或者满足LEV I排放标准的总质量为3 751-5 750磅的轻型卡车。

"LVM"指大型车企。

"MDV" 指中型车企。

"NMOG" 是指非甲烷有机气体，指被氧化和未被氧化的碳氢化合物的总质量。

"NEV" 指低速电动车。

"NOx" 指氮氧化物。

"PC" 是指乘用车。

"PZEV" 指部分配额零排放车，这些车辆指那些在加州交付销售，并有资格获得至少 0.2 分的配额积分的车辆。

"SAE" 是指（美国）汽车工程师协会。

"SULEV" 是指特超低排放车。

"TZEV" 是指过渡型零排放车。

"UDDS" 是指市区测功机驾驶循环。

"US" 是指美利坚合众国。

"US06" 是指 US06 补充联邦测试程序。

"VMT" 是指行驶里程。

"ZEV" 指零排放车。

（k）可分割性。本节中的每条规定都是可分割的，如本节的某一条规定被认定为无效，其余的所有规定仍然继续完全有效。

（l）公开披露。董事会拥有符合 1962.1 要求的车辆记录，这些记录应当作为公开记录进行如下披露：

（1）对于 2018 年及以后年款，每个车企的年产量和其从每辆 ZEV、TZEV 获取的相应积分。

（2）每家车企 2018 及以后年款的积分余额。

（A）每种车辆的类型：ZEV（不包括 NEV）、BEVx、NEV、TZEV 以及 AT PZEV 和 PZEV 的折扣积分；

（B）先进技术示范项目；

（C）交通系统；以及

（D）根据 1962.2（d）（5）（A）条款获取的积分，包括从第三方处获取或转移的积分。

备注：文献引用：《健康和安全法》第 39600、39601、43013、43018、43101、43104 和 43105 节。文献参考：《健康和安全法》第 38562、39002、39003、39667、43000、43009.5、43013、43018、43018.5、43100、43101、43101.5、43102、43104、43105、43106、43107、43204、和 43205.5 节。

§ 1962. 3 电动车充电要求

（a）适用性。本节适用于：

（1）除 2006 - 2013 年款低速电动车外，其它所有根据 1962.1 和 1962.2 的规定可获得 ZEV 积分的电动车、里程延长式电动车；

（2）能够通过电网连接电池充电器对车辆牵引电池进行充电的所有混合动力电动车；

（b）定义

（1）第 1962.1 和 1962.2 节中的定义适用于本节。

（c）要求

（1）从 2006 年款开始，（a）条款中确认的所有车辆必须配备一个导电充电器接口和充电系统，该充电系统符合所有适用于交流电一级和二级的充电要求，可参阅《汽车工程师协会（SAE）地面车辆推荐做法 SAE J1772》（2010 年 1 月版本）和《SAE 电动车和插电式混合动力电动车传导充电耦合器》。所有这些车辆也必须配备一个最小输出功率为 3.3 千瓦的车载充电器，或配备一个在 4 个小时之内能够使车辆完全荷电的车载充电器。

（2）车企可以向执行官员申请使用（c）（1）条款中规定的交流电插口的替代方案，但必须满足以下条件：

    （A）每个车辆应配备一个标准适配器，该适配器可使该车辆满足（c）（1）条款中规定的所有其他系统和车载充电器的要求；

    （B）标准适配器和插口替代方案必须由美国国家认可测试实验室（NRTL）检测认可。

备注：文献引用：《健康和安全法》第 39600、39601、43013、43018、43101、43104 和 43105 节。文献参考：《健康和安全法》第 38562、39002、39003、39667、43000、43009.5，43013、43018、43018.5、43100、43101、43101.5、43102、43104、43105、43106，43107，43204、43205.5 节。

2

加州低排放车法规（第三阶段）

§1961.2　2015 及后续年款乘用车、轻型卡车和中型车辆尾气排放标准和测试程序

简介：本节 1961.2 规定了加州"LEV III"乘用车（PC）、轻型卡车（LDT）和中型车辆（MDV）2015 及后续年款的尾气排放标准。车企必须证明遵守（a）款中适用于特定检测组的排放标准，并符合（b）款中适用于车企整体车队的综合渐进式目标。

2015 年款以前，生产符合（a）款标准车辆的车企可以选择对车辆进行标准认证，经认证的车辆将被视为 LEV III 车辆，适用车队渐进式目标。同样地，可以对 2015－2019 年款的车辆进行 1961（a）（1）款所述 LEV II 排放标准认证，经认证的车辆将被视为 LEV II 车辆，适用车队渐进式目标。

车企可以选择对车辆总重量（GVW）大于 10 000 磅（lbs.）的非完整 MDV 和柴油 MDV 使用的发动机进行 1956.8（c）和（h）款 CCR 第 13 主题所述的重型发动机标准和检测程序认证。所有小于或等于 10 000 lbs. GVW 的 MDV，包括非完整奥托循环 MDV 和使用柴油循环发动机的 MDV，都必须通过本 1961.2 规定的 2020 及后续年款的 LEV III 底盘标准和检测程序的认证。

条款

对于每个年款，车企应当证明在整个年款遵守本 1961.2 规定的标准，有以下两个途径可供选择：

途径 1：符合（a）款和 1961（a）（1）款所述加州尾气排放标准的、生产并在加州交付销售的 PC、LDT 和 MDV 的总数；或者

途径 2：符合（a）款和 1961（a）（1）款所述加州尾气排放标准的、生产并在加州、哥伦比亚地区和根据联邦清洁空气法（42 U.S.C.§7507）第 177 条的规定、采用本 1961.2 所述加州污染物排放标准的所有各州交付销售的 PC、LDT 和 MDV 的

总数。

选择途径2的车企应当在适用年款开始以前、以书面形式告知执行官员，否则就要遵守途径1。一旦车企选择途径2，那么该选择将生效，除非车企在适用年款开始以前以书面形式告知执行官员其将选择途径1。

如果车企某年款选择途径2，则1961.2所述的"在加州"指的是加州、哥伦比亚地区、以及根据联邦清洁空气法（42 U.S.C. § 7507）第177章的规定、采用1961.2所述加州污染物排放标准的所有各州。

（a）尾气排放标准

（1）"LEV III"尾气排放标准。下面所列标准是2015及后续年款LEV III PC、LDT和MDV在整个使用寿命期间内的最大尾气排放，包括使用气体或乙醇燃料的灵活燃料、双燃料及混合燃料车辆。2015 – 2019年款的LEV II LEV车辆可以进行本（a）（1）款所述适用LEV160、LEV395或LEV630的150 000英里耐久性非甲烷有机气体 + 氮氧化物（NMOG + NOx）排放联合标准认证，以及本（a）（4）款所述NMOG + NOx数值认证，来代替1961（a）（1）款所述单独的非甲烷有机气体和氮氧化物（NMOG和NOx）尾气排放标准和1961（a）（4）款所述NMOG数值认证；LEV II ULEV车辆可以进行本（a）（1）款所述适用ULEV125、ULEV340或ULEV570的150 000英里NMOG + NOx排放联合标准认证，以及（a）（4）款所述相应的NMOG + NOx数值认证，来代替1961（a）（1）款所述单独的NMOG和NOx尾气排放标准认证和1961（a）（4）款所述相应的NMOG数值认证。根据《加州2009 – 2017年款乘用车、轻型卡车和中型车辆的零排放车和混合动力电动车尾气排放标准》获得部分ZEV配额的2015 – 2019年款的LEV II SULEV车辆，和根据（b）（2）款的规定，允许进行利用"结转"排放检测数据来通过LEV II SULEV标准认证的2015 – 2016年款车辆，可以进行本（a）（1）款所述适用SULEV30、SULEV170或SULEV230的150 000英里NMOG + NOx排放联合标准认证，以及（a）（4）款所述相应的NMOG + NOx数值认证，以代替1961（a）（1）款所述单独的NMOG和NOx尾气排放标准认证和1961（a）（4）款所述相应的NMOG数值认证。如果LEV II SU-LEV车辆既没有获得部分ZEV配额，也没有通过2015 – 2016年款利用"结转"排放检测数据来通过LEV II SULEV标准认证，则不能进行NMOG + NOx联合标准认证。经NMOG + NOx联合标准认证的LEV II车辆将被视为LEV II车辆，适用车队渐进式目标。

PC、LDT 和 MDV2015 及后续年款 LEV III 尾气排放质量标准						
车型	车辆的耐久性基准（mi）	车辆排放种类[2]	NMOG + NOx（g/mi）	CO（g/mi）	HCHO（mg/mi）	颗粒物（g/mi）[1]
所有 PC；LDT 8 500 lbs. GVWR（磅.额定总重量）/或更小；MDPV 本类型的车辆在满载下检测。	150 000	LEV160	0.160	4.2	4	0.01
		ULEV125	0.125	2.1	4	0.01
		ULEV70	0.070	1.7	4	0.01
		ULEV50	0.050	1.7	4	0.01
		SULEV30	0.030	1.0	4	0.01
		SULEV20	0.020	1.0	4	0.01
MDV 8501 – 10 000 lbs. GVWR 本类型的车辆在校正的满载重量下检测。	150 000	LEV395	0.395	6.4	6	0.12
		ULEV340	0.340	6.4	6	0.06
		ULEV250	0.250	6.4	6	0.06
		ULEV200	0.200	4.2	6	0.06
		SULEV170	0.170	4.2	6	0.06
		SULEV150	0.150	3.2	6	0.06
MDV 10 001 – 14 000 lbs. GVWR 本类型的车辆在校正的满载重量下检测。	150 000	LEV630	0.630	7.3	6	0.12
		ULEV570	0.570	7.3	6	0.06
		ULEV400	0.400	7.3	6	0.06
		ULEV270	0.270	4.2	6	0.06
		SULEV230	0.230	4.2	6	0.06
		SULEV200	0.200	3.7	6	0.06

[1] 这些标准仅适用于未纳入（a）(2) 款所述颗粒物标准渐进式目标的车辆。

[2] 类型名称的数值部分为 NMOG + NOx 值，单位克/英里（g/mi），保留到 0.001 位。

（2）"LEV III" 颗粒物标准

（A）PC、LDT 和 MDV 颗粒物标准。从 2017 年款开始，除了小型车企，其他车企应当根据以下渐进式目标进度表，使其 PC、LDT 和 MDPV 车队一定比例的车辆通过以下颗粒物标准认证。这些标准是车辆使用寿命期限内的最大允许颗粒物排放量。所有进行上述颗粒物标准认证的车辆必须通过（a）(1) 款中所述的 LEV III 尾气排放标准认证。

PC、LDT 和 MDPVLEV III 颗粒物排放标准与渐进式目标		
年款	通过 3mg/mi 标准认证的车辆比例（%）	通过 1mg/mi 标准认证的车辆比例（%）
2017	10	0
2018	20	0
2019	40	0
2020	70	0
2021	100	0
2022	100	0
2023	100	0
2024	100	0
2025	75	25
2026	50	50
2027	25	75
2028 及以后	0	100

（B）MDV（不包括 MDPV）颗粒物标准。

1. 从 2017 年款开始，除了小型车企，其他车企应将其 MDV 车队一定比例的车辆通过以下颗粒物标准认证。这些标准为车辆使用寿命期限内的最大允许颗粒物悬浮颗粒排放量。所有进行上述颗粒物标准认证的车辆必须通过（a）（1）款中所述的 LEV III 尾气排放标准认证。本（a）（2）（B）1 款不适用于 MDPV。

MDV（不包括 MDPV）LEV III 颗粒物排放标准值	
车辆类型[1]	颗粒物（mg/mi）
MDV 8 501 – 10 000 lbs. GVWR，MDPV 除外	8
MDV 10 001 – 14 000 lbs. GVWR	10

[1] 本类型的车辆在校正后的满载重量下检测。

2. 除了小型车企，其他 MDV 车企应当根据以下渐进式目标进度表，证明其至少满足 MDV 车队中下述比例的车辆通过（a）（2）（B）1 款中规定的颗粒物标准认证。本（a）（2）（B）2 款不适用于 MDPV。

MDV（不包括MDPV）LEV III 颗粒物排放渐进式目标	
年款	根据适用性，通过 8mg/mi 颗粒物标准或者 10mg/mi 颗粒物标准认证的 MDV 总比例（%）
2017	10
2018	20
2019	40
2020	70
2021 及以后	100

（**C**）**小型车企颗粒物标准**。2021－2027 年款期间，小型车企应当将其 PC、LDT 和 MDPV 车队的车辆 100% 实现 3 mg/mi 颗粒物标准认证。2028 及以后的年款，小型车企应当将其 PC、LDT 和 MDPV 车队的车辆 100% 实现 1mg/mi 颗粒物标准认证。对 2021 及后续年款，小型车企应当将其除 MDPV 之外的 8501－10 000 lbs. GVWR MDV100% 实现 8mg/mi/英里颗粒物标准认证。对 2021 及后续年款，小型车企应当将其生产的 10 001－14 000 lbs. GVWR MDV100% 实现 10 mg/mi 的颗粒物标准认证。这些标准都是车辆使用寿命期限内的最大允许排放量。所有执行这些标准的车辆必须通过（a)(1）款中规定的 LEV III 尾气排放标准认证。

（**D**）**可选颗粒物标准渐进式进度表**。

1. PC、LDT 和 MDPV 的可选3mg/mi 颗粒物标准渐进式进度表。车企可以使用可选渐进式进度表，以达到 3 mg/mi 的颗粒物标准渐进式目标，只要 PC、LDT 和 MDPV 在 2021 年款实现等量的颗粒物减排。年款减排应当按照下列方式计算：将某年款中（基于车企预计的每个种类车辆的销售量），符合 3mg/mi 的颗粒物标准车辆的比例乘以该年款距实现渐进式目标的年数，即 2017 年款乘以 5，2018 年款乘以 4，2019 年款乘以 3，2020 年款乘以 2，2021 年款乘以 1。PC＋LDT＋MDPV 的年度计算结果应当加总得出该年 PC＋LDT＋MDPV 的减排总额。2021 年款的减排总额应当等于或者大于490，并且车企 100% 的 PC、LDT 和 MDPV 都应通过相当于 3mg/mi 的颗粒物标准认证。车企可以将 2017 年款之前生产的车辆计入减排总额计算公式（比如，2016 年款符合标准的车辆比例乘以 5）。

2. PC、LDT 和 MDPV 可选的 1 mg/mi 颗粒物标准渐进式进度表。车企可以采用可选渐进式进度表，以达到 1mg/mi 的颗粒物标准渐进式目标，只要 PC、LDT 和 MDPV 在 2028 年款实现等量的颗粒物减排。年款减排应当按照下列方式计算：将某年款中（基于车企预计的每个种类车辆的销售量），符合 1mg/mi 的颗粒物标准车辆的比例乘以该年款距实现渐进式目标的年数，即 2025 年款乘以 4，2026 年款乘以 3，2027 年款乘以 2，2028 年款乘以 1。PC + LDT + MDPV 年度计算结果应当加总，以得到 PC + LDT + MDPV 的减排总额。2028 年款的减排总额应当等于或者大于 500，并且车企 100% 的 PC、LDT 和 MDPV 都应通过与 1mg/mi 相当的的颗粒物标准认证。车企可以将 2025 年款之前生产的车辆计入减排总额计算公式（比如，2024 年款符合标准的车辆比例要乘以 4）。

3. MDV（不包括 MDPV）的可选渐进式目标进度表。车企可以使用可选渐进式目标进度表，以达到颗粒物标准渐进式目标，只要 MDV（不包括 MDPV）在 2021 年款实现等量的颗粒物减排。年款减排应当按照下列方式计算：将某年款中（基于车企预计的每个种类车辆的销售量），符合 8 mg/mi 或者 10 mg/mi 颗粒物标准的 MDV 的比例乘以该年款距实现渐进式目标的年数，即 2017 年款乘以 5，2018 年款乘以 4，2019 年款乘以 3，2020 年款乘以 2，2021 年款乘以 1。MDV 年度计算结果应当加总，以得到 MDV 的减排总额。2021 年款的减排总额应当等于或者大于 490，并且车企 100% 的 MDV 都应符合 8 mg/mi 或相当于 10 mg/mi 的颗粒物标准。车企可以将 2017 年款之前生产的车辆计入减排总额计算工时（比如，2016 年款的车辆比例要乘以 5）。

（3）双燃料（bi-fuel），灵活燃料（fuel-flexible）及混合燃料（dual-fuel）车辆 NMOG + NOx。对于灵活燃料、双燃料及混合燃料的 PC、LDT 和 MDV，NMOG + NOx 尾气排放标准合规性必须基于车辆燃烧气体或乙醇燃料时以及燃烧汽油时的尾气排放测试。当车企对车辆进行燃烧气体或乙醇燃料、汽油或柴油认证时，必须证明其符合（a）（1）款表格所示的 NMOG + NOx、一氧化碳（CO）和甲醛（HCHO）尾气排放标准。

当灵活燃料、双燃料及混合燃料车辆燃烧汽油时，车企可以根据《加州乘用车、轻型卡车和中型车辆 2015 及后续年款尾气排放标准和测试程序以及 2017 及后续年款温室气体排放标准和测试程序》测量 NMHC，替代原来的 NMOG。灵活燃料、双燃料及混合燃料车辆燃烧汽油时不要求进行 50 ℉ 的测试。

（4）50 ℉①尾气排放标准。所有 PC、LDT 和 MDV（非天然气和柴油动力车辆）必须被证明符合以下 NMOG + NOx 和 HCHO 尾气排放标准，这些标准通过 FTP（40 CFR，第 86 部分，B 子部分）在正常 50 ℉ 测试温度下衡量，并经《加州乘用车、轻型卡车和中型车辆 2015 及后续年款尾气排放标准和测试程序以及 2017 及后续年款温室气体排放标准和测试程序》第 II 部分，C 款修订。车企可以通过测量 NMHC 尾气排放证明对本段所述 NMOG + NOx 和 HCHO 认证标准的合规性，或者发布一个

---

① 50 ℉ = 10℃

声明，证明根据《加州乘用车、轻型卡车和中型车辆 2015 及后续年款尾气排放标准和测试程序以及 2017 及后续年款温室气体排放标准和测试程序》的 D.1 款（p）段和 G3.1.2 款对 HCHO 标准的合规性。50 ℉下测试的 CO 排放不能超过（a）(1) 款中针对同样排放种类的车辆以及需要在 68°①至 86°F② 下进行冷浸泡和排放测试的车辆的标准。

（A）经 LEV III 标准认证的 PC、LDT 和 MDPV 的排放标准。

LEV III PC、LDT 和 MDPV 50 ℉尾气排放标准			
车辆排放种类	NMOG + NOx（g/mi）		HCHO（g/mi）
	汽油	乙醇燃料	汽油和乙醇燃料
LEV160	0.320	0.320	0.030
ULEV125	0.250	0.250	0.016
ULEV70	0.140	0.250	0.016
ULEV50	0.100	0.140	0.016
SULEV30	0.060	0.125	0.008
SULEV20	0.040	0.075	0.008

（B）符合 LEV III 标准的 MDV（不包括 MDPV）的排放标准。

LEV III MDV（非 MDPV）50 ℉尾气排放标准			
车辆排放种类	NMOG + NOx（g/mi）		HCHO（g/mi）
	汽油	乙醇燃料	汽油和乙醇燃料
LEV395	0.790	0.790	0.064
ULEV340	0.680	0.680	0.032
ULEV250	0.500	0.500	0.032
ULEV200	0.400	0.500	0.016
SULEV170	0.340	0.425	0.016
SULEV150	0.300	0.375	0.016

---

① 68 ℉ = 20℃
② 86 ℉ = 30℃

LEV III MDV（非 MDPV）50 ℉尾气排放标准			
车辆排放种类	NMOG + NOx （g/mi）		HCHO （g/mi）
	汽油	乙醇燃料	汽油和乙醇燃料
LEV630	1. 260	1. 260	0. 080
ULEV570	1. 140	1. 140	0. 042
ULEV400	0. 800	0. 800	0. 042
ULEV270	0. 540	0. 675	0. 020
SULEV230	0. 460	0. 575	0. 020
SULEV200	0. 400	0. 500	0. 020

（5）低温 CO 标准。以下标准为 PC、LDT 和 MDPV 2015 及后续年款 50 000 英里低温下 CO 尾气排放水平。

PC、LDT 和 MDPV2015 及后续年款低温下 CO 尾气排放标准（g/mi）

车型	CO
所有 PC、LDT（0 – 3750 lbs. LVW	10. 0
LDT 3 751 lbs. LVW（装有负载的车辆总重，整备质量 + 300 磅）– 8 500 lbs. GVWR；MDPV 10 000 lbs. GVWR 或更小	12. 5

这些标准适用于 40 CFR，第 86 部分，B 子部分规定的（经《加州乘用车、轻型卡车和中型车辆2015 及后续年款尾气排放标准和测试程序以及 2017 及后续年款温室气体排放标准和测试程序》修订的）、正常 20 ℉（– 7℃）温度下测试的车辆。天然气、柴油动力和零排放车辆免除遵守这些标准。

（6）公路 NMOG + NOx 。根据联邦"公路燃料经济性测试"（HWFET, 40 CFR 600 B 子部分）测量（经《加州乘用车、轻型卡车和中型车辆 2015 及后续年款尾气排放标准和测试程序以及 2017 及后续年款温室气体排放标准和测试程序》修订）的 NMOG + NOx 最大排放不能超过（a）(1) 款中 LEV III NMOG + NOx。NMOG + NOx 排放量和 HWFET 标准在对比之前，都要根据 ASTM E29 – 67 就近保留至 0. 001 g/mi。

（7）补充联邦测试程序（SFTP）非使用状态排放标准。

　　（A）PC、LDT 和 MDPV SFTP NMOG + NOx 和 CO 排放标准。车企应当对 2015 及后续年款（PC、LDT 或 MDPV 类）的低排放型车辆（LEV）、超低排放型车辆（ULEV）和特超低排放型车辆（SULEV）进行（a）(7)(A) 1 款所述 SFTP NMOG + NOx 和 CO 独立尾气排放标准认证或证明符合（a）(7)(A) 2 款所述 SFTP

NMOG + NOx 和 CO 综合尾气排放标准和产品线平均目标。车企也可以对 PC、LDT 或 MDPV 类的 2014 年款的 LEV、ULEV 和 SULEV 进行 LEV III SFTP 标准认证，在这种情况下，车企应当遵守 LEV III SFTP 排放标准和要求，包括针对 2015 年款的按销售量加权后的产品线平均 NMOG + NOx 综合排放标准（如果车企选择遵守（a）(7)（A）2 款所述 SFTP NMOG + NOx 和 CO 综合尾气排放标准和产品线平均目标）。车企应当在 150 000 英里耐久性 SFTP NMOG + NOx 和 CO 排放标准认证的第一个测试组的认证申请表中，将其选择的排放标准类型告知执行官员。一旦为某车队选定 SFTP NMOG + NOx 和 CO 排放标准，并且执行官员已被告知该选择，则整个车队必须持续采用选定的标准至 2025 年款（如果选择采用（a）(7)（A）2 款的标准，该车队还包括 LEV II 车辆）。车企直到 2026 年款才能更改选择。不采用本（a）(7)（A）款所述 150 000 英里 SFTP NMOG + NOx 和 CO 排放标准认证的测试组应当符合 1960.1（r）款所述 4 000 英里 SFTP NMOG + NOx 和 CO 排放标准。

1. SFTP NMOG + NOx 和 CO 尾气单独排放标准。在进行 FTP 认证时，车辆使用同样的气体或者液体燃料，下列标准规定了 2015 及后续年款 LEV III LEVs，ULEVs 和 SULEVs 车辆在整个使用寿命期内 NMOG + NOx 和 CO 的最大排放量。针对灵活燃料的车辆，SFTP 合规需使用指定的 LEV III 认证汽油为燃料，可参见《加州乘用车、轻型卡车和中型车辆 2015 及后续年款尾气排放标准和测试程序以及 2017 及后续年款温室气体排放标准和测试程序》第 II 部分 A. 100. 3. 1. 2 款。

LEV III PC，LDT 和 MDPV 2015 及其后年款 SFTP NMOG + NOx 和 CO 尾气单独排放标准						
车型	车辆的耐久性基准（mi）	车辆排放种类[1]	US06 测试（g/mi）		SC03 测试（g/mi）	
			NMOG + NOx	CO	NMOG + NOx	CO
所有 PCs；LDTs 0 – 8 500 lbs. GVWR；和 MDPVs 这些分类下的车辆均是在满载下进行测试的。	150 000	LEV	0.140	9.6	0.100	3.2
		ULEV	0.120	9.6	0.070	3.2
		SULEV（选项 A）[2]	0.060	9.6	0.020	3.2
		SULEV	0.050	9.6	0.020	3.2

[1]**车辆排放种类**。通过了 150 000 英里耐久性 LEV III FTP 排放种类认证的所有车辆，车企必须证明它们满足上表所述的排放标准，或者更严格的 SFTP 排放种类标准。也就是说，所有通过了 150 000 英里 FTP 排放标准认证的 LEV III LEVs 车辆应当遵循表格中的 SFTP LEV 排放标准，所有通过了 150 000 英里 FTP 排放标准认证的 LEV III ULEVs 车辆应当遵循表格中的 SFTP ULEV 排放

标准，所有通过了 150 000 英里 FTP 排放标准认证的 LEV III SULEVs 车辆应当遵循表格中的 SFTP SULEV 排放标准。

²**可选 SFTP SULEV 标准**。车企要保证重量在 6 001 到 8 500 lbs. GVWR 之间的 LDT 测试组和 MDPV 测试组符合上述表格中针对 2015 – 2020 年款 SULEV 的排放标准，前提是只有当测试组车辆配备了微粒过滤器并且车企将微粒过滤器排放保修里程延长为 200 000 英里。重量在 0 – 6 000 lbs. GVWR 之间的 PC 和 LDT 不适于这一选项。

2. SFTP NMOG + NOx 和 CO 综合尾气排放标准。对于 2015 及后续年款，选择这一选项的车企必须保证 LEV II 以及 LEV III 的 LEV，ULEV 和 SULEV NMOG + NOx 按销售量加权的车队平均的综合排放值不超过下表中适用的 NMOG + NOx 综合气体排放标准。此外，任何 LEV III 测试组的 CO 综合排放值不得超过下表中 CO 综合排放标准。应该使用与 FTP 认证相同的气体或液体燃料来证明 SFTP 的合规性。针对灵活燃料车辆，需用 LEV III 认证的汽油来证明 SFTP 的合规性，可参见《加州 2015 及后续年款乘用车，轻型卡车和中型车辆标准污染物排放标准和测试程序以及 2017 及后续年款温室气体排放标准和测试程序》第 II 部分 A 节 . 100. 3. 1. 2 款。

对于本款中每个测试组，车企应当计算 NMOG + NOx 的综合排放值，并且针对 LEV III 测试组，用下列方程来单独计算 CO 综合排放值：

$$综合排放值 = 0.28 \times US06 + 0.37 \times SC03 + 0.35 \times FTP \qquad [方程 1]$$

其中 "US06" = 测试组的 NMOG + NOx 或 CO 排放值，如果适用的话，通过 US06 测试确定；

"SC03" = 测试组的 NMOG + NOx 或 CO 排放值，如果适用的话，通过 SC03 测试确定；

"FTP" = 测试组的 NMOG + NOx 或 CO 排放值，如果适用的话，通过 FTP 试验确定；

如果测试组的车辆都没有空调，FTP 测试循环排放值可以用来代替方程 1 中 SC03 测试循环排放值。为了确定是否遵守适用于该年款的 SFTP NMOG + NOx 综合排放标准，车企应当对每一个可适用的测试组，使用在销售量加权下的的 NMOG + NOx 综合排放量的车队平均值。销售量加权的车队平均值应当由结转额和新认证的 SFTP 综合排放量（转换成 NMOG + NOx，如适用的话）联合计算得出。LEV II 测试组在计算车队平均值时将使用它们的排放值，但在 LEV III 测试组中将不会考虑。

是否符合 CO 综合排放标准不能通过车队平均值来验证。NMOG + NOx 销售量加权的车队平均综合排放值以及每个测试组的 CO 综合排放值不得超过下表中的值：

2015 及后续年款 PC，LDT 和 MDPV SFTP NMOG + NOx 和 CO 综合排放标准（g/mi）[1]										
年款	2015	2016	2017	2018	2019	2020	2021	2022	2023	2024	2025 +
所有的 PCs、LDTs ≤8 500 lbs. GVWR 和 MDPVs[3]. 该车型的车辆均是在满载下进行测试（即整备质量 + 300 磅），LEV II 车辆除外（其测重参考§1960.1（r）第 13 编，CCR。	销售量加权车队平均的 NMOG + NOx 综合尾气排放标准[2456]									
	0.140	0.110	0.103	0.097	0.090	0.083	0.077	0.070	0.063	0.057	0.050
	CO 综合尾气排放标准[7]									
	4.2									

[1] **合规测试里程数**。所有符合 150 000 英里条件下 LEV III FTP 排放标准的测试组，应当按照《加州乘用车，轻型卡车和中型车辆 2015 及后续年款标准污染物排放标准和测试程序以及 2017 及后续年款温室气体排放标准和测试程序》的规定进行 150 000 英里的 SFTP 测试。

[2] **确定 LEV II 测试组和联邦清洁车辆的 NMOG + NOx 综合排放值**。对于通过了 LEV II FTP 排放标准认证的测试组，SFTP 排放值将转换成 NMOG + NOx 并且在使用劣化系数或老化组件条件下预计达到 120 000 或者 150 000 英里（取决于 LEV II FTP 认证）。为了确定结转综合排放值以计算 NMOG + NOx 销售量加权的车队平均值，LEV II 测试组可不使用 SFTP 测试循环的劣化系数，而使用从 FTP 测试循环获得的适用劣化系数。如果 SFTP 整个使用寿命期间的排放值用来进行 LEV II SFTP 4k 标准认证，那么此排放值可在销售量加权的车队平均值计算中使用，不再额外考虑劣化系数。对于通过了联邦认证的测试组，在加州根据《加州乘用车，轻型卡车和中型车辆 2015 及后续年款标准污染物排放标准和测试程序以及 2017 及后续年款温室气体排放标准和测试程序》第 H.1.4 款规定进行核证时，用来满足联邦 SFTP 排放要求的数值，可被用在销售量加权的车队平均值计算中，而不再考虑劣化系数。在所有情况下，US06 和 SC03 测试循环的 NMHC 排放值乘以系数 1.03 可以转换成 NMOG 排放值。

[3] MDPVs 只有通过 LEV III 150 000 英里 NMOG + NOx 和 CO 排放要求认证才适用于 SFTP NMOG + NOx 和 CO 排放标准以及销售量加权的车队平均值。

[4] LEV III 测试组必须保证排放量本值在 0.010 g/mi 增幅以内。以 2018 年款车型为起点，车辆本值排放量不得超过 0.180 g/mi 的最大增幅。

[5] **计算 NMOG + NOx 销售量加权的车队平均排放量**。针对每一年款，车企应当按照下列公式来计算销售量加权的车队平均的 NMOG + NOx 综合排放值。

$$\frac{\sum_{i=1}^{n}(测试组车辆数)_i \times (复合本值)_i}{\sum_{i=1}^{n}(测试组车辆数)_i} \qquad [方程.2]$$

其中 "n" = 车企的 PC, LDT 总数量，如果适用，还包括 MDPV 本值（包括结转认证值在内，符合此年款 SFTP 综合排放标准）。

"测试组车辆数量" = 认证测试组的车辆数量，这些车辆是生产并在加州交付销售的。

"综合本值" = 车企选择的认证本值，可作为进行 SFTP 150 000 英里耐久性所有测试的测试组车辆的排放标准，以及 LEV II 测试组的 SFTP 结转排放值（参照表格的脚注2）。

[6]**计算平均 NMOG + NOx 借记与积分**。车企可以依照下列公式计算总 NMOG + NOx 借记与积分：

[（NMOG + NOx 综合排放标准） − （（车企销售量加权的车队平均综合排放值）]（生产并在加州交付销售的车辆总数，这些车辆类别均为 0 − 8 500lbsGVW + MDPVs） [方程.3]

结果是负数，则 NMOG + NO 为借记总额，结果是正数，则 NMOG + NOx 为积分，结果计入车企指定的年款。某一年款获得的总 NMOG + NOx 积分在获得的五年内始终全部有效。从第六个年款开始，总 NMOG + NOx 积分作废。车企可以与其他车企进行积分交易。

车企可通过获得与 NMOG + NOx 借记额相等的积分，以在借记发生后三个年款内偿清 NMOG + NOx 借记额。如果 NMOG + NOx 借记总额在三个年款期间未能偿清，车企将受到《健康与安全法》第 43211 节规定的民事处罚。当 NMOG + NOx 借记总额在指定期限期满时未能偿清，则构成处罚的理由。根据《健康和安全法》第 43211 条款，不符合州理事会采用的排放标准的车辆数，通过发生借记总额的年款期间 NMOG + NOx 综合排放标准除以此该年款期间的 NMOG + NOx 借记总额来确定。

[7]**计算 CO 综合排放值**。CO 综合排放值应当依据上文的方程 1 来计算。与 NMOG + NOx 综合排放标准不同，车企达到车队平均标准来并不能通过 CO 综合排放标准认证：每个测试组都必须符合标准。通过了 4 000 英里 SFTP 排放标准认证的测试组不受 CO 排放标准的约束。

（B）PC，LDT 和 MDPV 的 SFTP 颗粒物尾气排放标准。当测试组使用与 FTP 测试相同的气体或液体燃料时，下列标准规定了 2017 及后续年款 PC, LDT, 和 MDPV 车型中 LEV III 的 LEV, ULEV 和 SULEV 在整个使用寿命期间 PM 尾气最大排放值。对于灵活燃料车辆，SFTP 合规需使用指定的 LEV III 认证汽油为燃料，可参见《加州乘用车、轻型卡车和中型车辆 2015 及后续年款尾气排放标准和测试程序以及 2017 及后续年款温室气体排放标准和测试程序》第 II 部分第 A. 100. 3. 1. 2 款。车企必须证明 PC, LDT 和 MDPV 车型中 LEV III 的 LEV, ULEV 和 SULEVs 在满足了第（a）(2) 条款中 LEV III FTP PM 150 000 英里排放标准的条件下，符合第（a）(7)（B）条款中 SFTP PM 尾气排放标准。

LEV III PC，LDT 和 MDPV 2017 及后续年款 SFTP 颗粒物尾气排放标准[1]				
车型	测重	合规测试里程	测试循环	颗粒物（mg/mi）
所有的 PC； LDTs0 – 6 000 lbs GVWR	满载车重	150 000	US06	10
LDTs 6 001 – 8 500 lbsGVWR； MDPVs	满载车重	150 000	US06	20

[1] 所有满足了（a)(2) 条款中 LEV III FTP PM 150 000 英里耐久性排放标准的 PC，LDT 和 MDPV 应遵守表格中 SFTP PM 尾气排放标准。

（C）MDV 的 SFTP NMOG + NOx 和 CO 尾气排放标准。当测试组使用与 FTP 认证相同的气体或液体燃料时，下列标准规定了 2016 及后续年款（总重在 8 501 – 14 000 磅之间）中型 LEV III ULEVs 和 SULEVs 在整个使用期间 NMOG + NOx 和 CO 最大综合排放量。对于灵活燃料车辆，SFTP 需使用指定的 LEV III 认证汽油 为燃料，可见《加州乘用车、轻型卡车和中型车辆 2015 及后续年款尾气排放标准 和测试程序以及 2017 及后续年款温室气体排放标准和测试程序》第 II 部分 A.100.3.1.2 条款。受（a)(7)(A) 条款和（a)(7)(B) 条款中排放标准约束的 MDPVs 不适用下列综合排放标准。

2016 及后续年款中型 ULEV 和 SULEV SFTP NMOG + NOx 和 CO 综合尾气排放标准						
车型	合规测试里程	HP/GVWR[2]	测试循环[345]	车辆排放种类[6]	综合排放标准（g/mi）[1]	
					NMOG + NOx	CO
MDV 8 501 – 10 000 lbs GVWR	150 000	≤0.024	US06 Bag 2，SC03，FTP	ULEV	0.550	22.0
				SULEV	0.350	12.0
		>0.024	Full US06，SC03，FTP	ULEV	0.800	22.0
				SULEV	0.450	12.0
MDV10 001 – 14 000 lbs GVWR	150 000	n/a	Hot 1435 UC（Hot 1435 LA92），SC03，FTP	ULEV	0.550	6.0
				SULEV	0.350	4.0

[1] 对于每一个测试组，车企可使用（a)(7)(A) 2 条款中方程 1 来计算它们的 SFTP 综合排放 值。对于车重（GVWR）在 10 001 – 14 000 磅之间的 MDV，UC 测试的排放结果将代替 US06 测试 的结果。

²**功重比**。如果测试组中所有车辆的功重比等于或小于 0.024，将对测试组车辆进行 US06 Bag 2 测试，而不是完整的 US06 测试循环。这个临界值由发动机的最大额定功率（由发动机厂商在申请认证时确定的），与车辆的额定车重之比来确定，对于混合动力电动汽车和插电式混合动力电动汽车，则不包括电机的功率。车企也可以选择进行全循环测试，而不考虑计算出来的比值，在这种情况下，车企应当符合适用于功重比大于 0.024 的车辆的排放标准。

³**测试车重（测重）**。MDV 在校正后的满载车重（汽车整备质量和额定车重的平均值）条件下进行测试。

⁴**变速型风扇**。车企在 MDV SFTP 测试中可以选择使用变速型风扇 – 参照 40 – CFR § 86. 107 – 96（d）(1) 条款中的说明，来代替定速型风扇。

⁵如果车企提供一份工程评估，说明测试组的 SC03 排放等于或小于 FTP 排放，则 FTP 排放值将代替 SC03 排放值来确定测试组综合排放值。

⁶**车辆排放种类**。对于额定车重在 8 501 – 10 000 lbs. 的 MDV，每一年款间，符合 1961.2 条款所述 SFTP 排放种类认证的 MDVs 比例应当等于或大于符合 FTP ULEV250、ULEV200、SULEV170、和 SULEV150 排放种类认证的总比例，这些车辆中，经 SFTP SULEV 排放种类认证的 MDV 比例应当等于或大于同时经 FTP SULEV170 和 SULEV150 排放种类认证的总比例。对于额定车重在 10 001 – 14 000 lbs. 之间的 MDVs，每一年款间，经 1961.2 条款所述 SFTP 排放种类认证的 MDVs 比例应当等于或大于经 FTP ULEV400、ULEV270、SULEV230 和 SULEV200 排放种类认证的总比例，这些车辆中，经 SFTP SULEV 排放种类认证的 MDV 比例应当等于或大于同时经 FTP SULEV230 和 SULEV200 排放种类认证的总比例。

（D）MDV SFTP 颗粒物尾气排放标准。当测试组使用与 FTP 认证相同的气体或液体燃料时，下列标准规定了 2017 及后续年款 LEV III 的 LEV，ULEV 和 SULEV 在其整个使用期间 PM 最大综合排放量。对于灵活燃料车辆，SFTP 合规需用指定的 LEV III 认证汽油为燃料，这在《加州乘用车、轻型卡车和中型车辆 2015 及后续年款尾气排放标准和测试程序以及 2017 及后续年款温室气体排放标准和测试程序》第 II 部分第 A. 100. 3. 1. 2 款。采用（a）(7)(A) 条款和（a）(7)(B) 条款中排放标准的 MDPVs 不适用下列综合排放标准。

MDV2017 及以后年款 SFTP 颗粒物尾气排放标准¹					
车型	测重	合规测试里程	Hp/GVWR²	测试循环³⁴	PM（mg/mi）
MDV 8 501 – 10 000 lbs GVWR	校正满载车重	150 000	≤ 0.024	US06 Bag 2	7
			> 0.024	US06	10
MDV10 001 – 14 000 lbs GVWR	校正满载车重	150 000	n/a	Hot 1435 UC（Hot 1435 LA92）	7

¹除采用（a）(7)(B) 条款规定的排放标准的 MDPV 外，通过了（a）(2) 条款中 150 000 英里耐久性 FTP PM 排放标准认证的 MDVs 应当遵守上述表格中 SFTP 颗粒物尾气排放标准。

²**功重比**。如果测试组中所有车辆的功重比等于或小于 0.024，将对测试组车辆进行 US06 Bag 2 测试，而不是完整的 US06 测试循环。这个临界值由发动机的马力值与车辆的额定车重磅值之比来确定，对于混合动力电动汽车和插电式混合动力电动汽车，该马力值不包括电机的马力。车企

也可以选择进行全循环测试，而不考虑计算出来的比值，在这种情况下，车企应当符合功重比大于 0.024 的车辆所适用的排放标准。

[3]**变速型风扇**。车企在 MDV SFTP 测试中可以选择使用变速型风扇 – 参照 40 – CFR § 86.107 – 96（d）(1) 条款中的说明，来代替定速型风扇。

[4]对于采用上表排放标准的每一个测试组，车企使用上文方程 1 来计算 SFTP PM 综合排放量。对于通过了 US06 Bag 2 PM 排放标准认证且额定车重（GVWR）在 8 501 - 10 000 磅之间的 MDV，US06 Bag 2 测试的排放结果将代替 US06 测试的结果。对于 GVWR 在 10 0001 - 14 000 之间的 MDV，UC 测试的排放结果将代替 US06 测试的结果。

（8）临时在用合规标准。

（A）LEV III NMOG + NOx 临时在用合规标准。下列临时在用合规标准适用于测试组通过 LEV III 标准认证后的前两年年款。

1. PC，LDT 和 MDPV NMOG + NOx 临时在用合规标准。这些标准适用于 2015 - 2019 年款。

排放种类	车辆的耐久性基准（mi）	LEV III PC, LDT 和 MDPV NMOG + NOx （g/mi）
LEV160	150 000	不适用
ULEV125	150 000	不适用
ULEV70	150 000	0.098
ULEV50	150 000	0.070
SULEV30	150 000	0.042[1]
SULEV20	150 000	0.028[1]

[1]不适用于获得 PZEV 积分的测试组

2. MDPV 除外的 MDV NMOG + NOx 临时在用合规标准。这些标准适用于 2015 - 2020 年款。

排放种类	车辆的耐久性基准（mi）	LEV III MDV（不包括 MDPV）8501 – 10 000lbs. GVW NMOG + NOx（g/mi）	LEV III MDV 10 001 – 14 000 lbs. GVW NMOG + NOx（g/mi）
LEV395	150 000	不适用	不适用
ULEV340	150 000	不适用	不适用
ULEV250	150 000	0.370	不适用

排放种类	车辆的耐久性基准（mi）	LEV III MDV（不包括 MDPV）8501 – 10 000lbs. GVW	LEV III MDV 10 001 – 14 000 lbs. GVW
		NMOG + NOx（g/mi）	NMOG + NOx（g/mi）
ULEV200	150 000	0.300	不适用
SULEV170	150 000	0.250	不适用
SULEV150	150 000	0.220	不适用
LEV630	150 000	不适用	不适用
ULEV570	150 000	不适用	不适用
ULEV400	150 000	不适用	0.600
ULEV270	150 000	不适用	0.400
SULEV230	150 000	不适用	0.340
SULEV200	150 000	不适用	0.300

（B）LEV III 颗粒物临时在用合规标准。下列临时在用合规标准适用于测试组通过 LEV III 标准认证后的前两年年款。

1. PC，LDT 和 MDPV 的 LEV III 颗粒物临时在用合规标准。对于 2017 – 2020 年款，通过了 3mg/mi 微粒标准认证的车辆，临时在用合规标准是 6mg/mi。对于 2025 – 2028 年款，通过了 1 mg/mi 颗粒物标准认证的车辆，临时在用合规标准是 2mg/mi。

2. MDV 的（MDPV 除外）LEV III 颗粒物临时在用合规标准。对于 2017 – 2020 年款，通过了 8mg/mi 颗粒标准认证的车辆，临时在用合规标准是 16mg/mi，通过了 10mg/mi 颗粒物标准认证的车辆，临时在用合规标准是 20 mg/mi。

（C）SFTP 临时在用合规标准。

1. 在 2020 年款之前认证的测试组在通过新标准认证的前两年年款可以使用 NMOG + NOx 在用合规标准。

a. 对于通过了（a）（7）（A）1 条款标准认证的 LDT（轻型车）测试组和 MDPV 测试组，NMOG + NOx 在用合规排放标准将是适用认证标准的 1.4 倍。

b. 对于通过了（a）（7）（A）2 条款标准认证的 LDT 测试组和 MDPV 测试组，NMOG + NOx 在用合规排放标准将是测试组认证的综合本值的 1.4 倍。

c. 对于通过了（a）（7）（C）条款标准认证的 MDV 测试组，NMOG + NOx 在

用合规排放标准将是适用认证标准的 1.4 倍。

2. 在 2021 年款之前认证的测试组在通过 SFTP 颗粒物标准认证的前五年年款可以使用颗粒物在用合规标准。

a. 对于通过了（a）（7）（B）条款中 SFTP 颗粒物尾气排放标准认证的 LDT 测试组和 MDPV 测试组，颗粒物在用合规排放标准将比适用认证标准高 5.0mg/mi。

b. 对于通过了（a）（7）（D）条款中 SFTP 颗粒物尾气排放标准认证的 MDV 测试组，颗粒物在用合规排放标准将比适用认证标准高 5.0mg/mi。

（9）产生额外的 NMOG + NOx 车队平均积分目标。对于一辆通过了（a）（1）条款 LEV III 标准认证的车辆，依据《加州乘用车、轻型卡车和中型车辆 2009 – 2017 年款零排放汽车、混合动力电动汽车的尾气排放标准和测试程序》和《加州乘用车、轻型卡车和中型车辆 2018 及后续年款零排放和混合动力电动车尾气排放标准和测试程序》的第 C.3 款规定，没有产生部分零排放车（PZEV）配额，如果车企将车辆的性能和缺陷保修期延长至 15 年或 15 万英里（以先达到者为准），零排放能源存储设备（如电池、超级电容或其他电能存储设备）发动的车辆保修期为 10 年，当计算车企的车队平均排放时，车企可以从（b）（1）（B）1.c 条款规定的 NMOG + NO 排放值中减去 5mg/mi。

（10）产生 PZEV 配额的目标。2015 – 2017 年款间，依据《加州乘用车、轻型卡车和中型车辆 2009 – 2017 年款零排放汽车和混合动力电动车尾气排放标准和测试程序》第 C.3 款规定的标准，符合了 LEV III SULEV30 或 LEV III SULEV20 标准的车企也产生了 PZEV 配额。

（11）直接减少臭氧技术的 NMOG 积分。如果保证车辆配备直接减少臭氧技术，车企将会得到 NMOG 积分，这些积分可以用于车辆达到 NMOG 尾气排放标准。为了得到积分，车企必须提交每一款车型的以下信息，为了它们能获得积分，必须提供包括但不限于以下资料：

（A）直接减少臭氧设备气流速度的演示以及设备在速度范围内减少臭氧的效率演示，速度的范围可参见统一测试循环驾驶附表，此附表包括在《加州乘用车、轻型卡车和中型车辆 2015 及后续年款标准污染物排放标准和测试程序以及 2017 及后续年款温室气体排放标准和测试程序》第 II G. 部分（2012 年 3 月 22 号通过）。

（B）车辆在整个使用寿命期间设备的耐久性评估；

（C）用来检测设备使用性能的车载诊断方案说明书。

依据以上资料，执行官员将使用一个核准的气流量模型，根据一个小时内臭氧峰值水平的计算变化来判定 NMOG 积分值。

（12）获得联邦认证的车型在加州的认证标准。

（A）基本目标。 如果车企将其 2015 及后续年款 PC，LDT 以及 MDV 车型获得了联邦特定排放本值标准认证（比适用的加州排放种类标准更加严格），相应的加州车型只需要进行以下标准认证：（i）加州车辆排放种类加州标准（至少和相应的联邦排放本值标准一样严格），或（ii）用来对联邦车型进行认证的尾气排放标准。然而，特定排放本值的联邦尾气排放标准与车辆排放种类的加州标准一样严格时，加州车型只需要被证明符合车辆排放种类的加州标准或者更加严格的加州标准。联邦排放本值包含在 40 CFR § 86. 1811 – 04（c）条款中表格 S04 – 1 和表格 S04 – 2 中，该条款于 2000 年 2 月 10 号通过。申请此目标的条件规定可参见《加州乘用车、轻型卡车和中型车辆 2015 及后续年款标准污染物排放标准和测试程序以及 2017 及后续年款温室气体排放标准和测试程序》第 I. H. 1 款。

（B）清洁燃料车辆除外。 若联邦认证的车型只销售给根据联邦《清洁空气法》（42 U. S. C. sec. 7586）第 246 条的规定符合清洁燃料车要求的运营商，则在加州申请认证时第（a）（12）（A）条款不适用。此外，针对联邦认证车型，当车企充分证明该车型主要出售给或租给清洁燃料车运营商使用，以及偶尔经推广销售或租赁给这些清洁燃料产品运营商，执行官员可以决定上述基本目标不适用该车型。

（13）燃油加热器的排放标准。 如果车企选择在 PC、LDT 和 MDV 上使用车载燃油加热器，该车载燃油加热器必须符合 1961（a）（1）条款规定的 ULEV125 标准，此标准适用于 GVWR 低于 8 500 磅的 PC 和 LDT。燃油加热器的尾气排放应当依据《加州乘用车、轻型卡车和中型车辆 2009 – 2017 年款零排放车和混合动力电动车尾气排放标准和测试程序》和《加州乘用车、轻型卡车和中型车辆 2018 及后续年款零排放车和混合动力电动车尾气排放标准和测试程序》（如果适用）来确定。如果车载燃油加热器能够在高于 40°F 的环境温度下正常工作，车载燃油加热器测试的排放水平将被增加到 FTP（40 CFR，第 86 部分，B 子部分）标准测试的排放值中，来判定是否符合（a）（1）条款的尾气排放标准，FTP 测试可参见《加州乘用车、轻型卡车和中型车辆 2015 及后续年款尾气排放标准和测试程序以及 2017 及后续年款温室气体排放标准和测试程序》。

（b）车企的排放标准渐进式目标。

（1）PC，LDT 和 MDPV 的 NMOG + NOx 车队平均排放目标。

（A） 由非小型车企生产并在加州交付销售的 PC，LDT 和 MDPV，其 NMOG + NOx 车队平均尾气排放值不得超过下表的规定：

年款	PC，LDT 和 MDPV 的 NMOG + NOx 车队平均尾气排放目标 (150 000 英里耐久性基准) NMOG + NOx 车队平均排放（g/mi）	
	所有的 PC；LDT 0 – 3750 lbs. LVW	LDT 3 751 lbs. LVW – 8 500 lbs. GVWR；所有的 MDPV
2014[1]	0.107	0.128
2015	0.100	0.119
2016	0.093	0.110
2017	0.086	0.101
2018	0.079	0.092
2019	0.072	0.083
2020	0.065	0.074
2021	0.058	0.065
2022	0.051	0.056
2023	0.044	0.047
2024	0.037	0.038
2025 +	0.030	0.030

[1]对于 2014 年款，车企可选择遵守上表格中 NMOG + NOx 车队平均排放值，而不是 1961（a）（b）(1)(A) 条款规定的 NMOG 车队平均排放值。车企必须遵守 PC/LDT1 车队和 LDT2/MDPV 车队的 NMOG + NOx 车队平均排放目标或者遵守 PC/LDT1 车队和 LDT2 车队的 NMOG 车队平均排放目标。车企必须使用适用的使用寿命标准来计算 NMOG + NOx 车队平均排放值。

1. 选择合规途径 2 的车企必须向执行官员提供在哥伦比亚特区，以及第 177 章所列各州生产并交付销售的每个测试组的车辆数目。

2. PZEV 防倒退目标。2018 及后续年款，车企必须保证生产并在加州交付销售最低比例的经 SULEV30 和 SULEV20 标准认证的 PC 和 LDT。对该车企来说，2015 – 2017 年款间，此最小比例必须与生产并在加州销售的 PZEVs 平均比例相等。车企使用这些年款预测的销售量代替实际销售量来计算平均比例。自 2020 年款起，经 SULEV30 和 SULEV20 标准认证的 PC 和 LDT 车队在适用年款和前两年年款的车队平均比例，将被用来判定是否符合该目标。

（B）计算 NMOG + NOx 车队平均排放量。

1. 基本计算。

a. 每一个车企所有生产并在加州交付销售的 PC 和 LDT1 NMOG + NOx 车队平均排放量计算公式如下：

（$\sum$［测试组的车辆数量,不包括具备车辆外充电能力的混合动力电动车 × 适用排放标准］+ $\sum$［测试组具备车辆外充电能力的混合动力电动车数量 × HEV NMOG + NOx 贡献系数］）÷ 生产并在加州交付销售的所有PC 和LDT1, 包括ZEV 和 HEV

b. 每一个车企所有生产并在加州交付销售的 LDT2 和 MDPV 的 NMOG + NOx 车队平均排放量计算公式如下:

（$\sum$［测试组的车辆数量,不包括具备车辆外充电能力的混合动力电动车 ×适用排放标准］+ $\sum$［测试组具备车辆外充电能力的混合动力电动车数量 × HEV NMOG 系数］）÷生产并在加州交付销售的所有 LDT2 和 MDPV,包括 ZEV 和 HEV

c. 上述公式中使用的适用排放标准如下:

年款	车型排放种类	排放标准值[1]（g/mi）	
		所有的 PC；LDT 0 - 3750 lbs. LVW	LDT 3 751 - 5 750 lbs. LVW；所有的 MDPV
联邦认证的 2015 及后续年款	所有	对车辆进行认证的 NMOG + NO$_x$ 联邦排放标准全使用寿命排放总量	对车辆进行认证的 NMOG + NO$_x$ 联邦排放标准全使用寿命排放总量
年款	车辆排放种类	所有的 PC；LDT 0 - 3750 lbs. LVW	LDT3 751 - 8 500 lbs. GVWR、所有的 MDPV
2015 - 2019 年款符合 1961（a）（1）条款中 "LEV II" 排放标准的车辆；2015 及后续年款符合 1961.2（a）（1）条款中 "LEV III" 排放标准的车辆；	LEV II LEV、LEV160	0.160	0.160
	LEV II ULEV、LEV125	0.125	0.125
	ULEV70	0.070	0.070
	ULEV50	0.050	0.050
	LEV II SULEV、SULEV30	0.030	0.030
	SULEV20	0.020	0.020
	LEV II LEV、LEV395	不适用	0.395

续表

年款	车型排放种类	排放标准值[1]（g/mi）	
		所有的 PC；LDT 0 – 3750 lbs. LVW	LDT 3 751 – 5 750 lbs. LVW；所有的 MDPV
2015 – 2019 年款符合 1961（a）（1）条款中"LEV II"排放标准的车辆； 2015 及后续年款符合 1961.2（a）（1）条款中"LEV III"排放标准的车辆；	LEV II ULEV	不适用	0.343
	ULEV340	不适用	0.340
	ULEV250	不适用	0.250
	ULEV200	不适用	0.200
	SULEV170	不适用	0.170
	SULEV150	不适用	0.150

[1] 符合第（a）(9) 条款所规定的延长保修期要求的 LEV III 型车辆测试组，适用排放标准值为上表中所列值减去 5 mg/mi。

2. 车辆外充电能力 HEV NMOG + NO$_x$ 贡献系数。轻型外充电混合动力电动车的 HEV NMOG + NO$_x$ 贡献系数计算方式如下：

LEV160 HEV 贡献系数 = 0.160 –［零排放 VMT 配额×0.035］

ULEV125 HEV 贡献系数 = 0.125 –［零排放 VMT 配额×0.055］

ULEV70 HEV 贡献系数 = 0.070 –［零排放 VMT 配额×0.020］

ULEV50 HEV 贡献系数 = 0.050 –［零排放 VMT 配额×0.020］

SULEV30 HEV 贡献系数 = 0.030 –［零排放 VMT 配额×0.010］

SULEV20 HEV 贡献系数 = 0.020 –［零排放 VMT 配额×0.020］

具备车辆外充电能力的充电混合动力电动车零排放 VMT 配额的确定须以《加州乘用车、轻型卡车和中型车辆 2009 – 2017 年款零排放车和混合动力电动车尾气排放标准和测试程序》与《加州乘用车、轻型卡车和中型车辆 2018 及后续年款零排放车和混合动力电动车尾气排放标准和测试程序》第 C.3 节中的适用性规定为准；此外，以上等式中最大许可零排放 VMT 配额为 1.0。第（b）(1)（B）2 条款只适用于符合（a）(1) 条款中"LEV III"排放标准的具备车辆外充电能力的 HEV。

（C）小型车企渐进式目标

1. 2015 – 2021 年款内，根据（b）(1)（B）条款计算，由小型车企所生产的所有 PC 以及 0 – 3750lbs. LVW 的 LDT 或是 3 751 – 5 750 lbs. LVW 的 LDT，车队平均 NMOG + NOx 值不得超过 0.160g/mi；2022 – 2024 年款内，由小型车企所生产的所有 PC、0 – 3750 lbs. LVW 的 LDT 或是 3 751 lbs. LVW – 8 500 lbs. GVW 的 LDT 及所有的 MDPV，车队平均 NMOG + NOx 值不得超过 0.125 g/mi；2015 及后续年款内，由小型车

企所生产的所有 PC、0 – 3750 lbs. LVW 的 LDT 或是 3 751 lbs. LVW – 8 500 lbs. GVW 的 LDT 及所有的 MDPV，车队平均 NMOG + NO$_x$ 值不得超过 0. 070 g/mi。2015 – 2021 年款内，小型车企可以对所产车辆进行 1961 节中的"LEV II"排放标准认证；2022 及后续年款内，小型车企的所有车辆都须符合 1961. 2 节中的"LEV III"排放标准。

2. 若某一车企先前连续三年在加州新产 PC、LDT、MDV、重型车辆（HDV）及重型发动机的平均销售量超出 4500（辆/台），这类车企将不再被视为小型车企。若此为该车企销量首次超过 4500，则应于上述连续三年之后的第四年开始，履行（b）(1)（A）条款中规定的大型车企车队平均排放目标；若在四年提前期内，该车企的销售量降至 4500 以下，之后又增至 4500 以上，则以该车企销售量再次超过 4500 的年份计为四年提前期的初始年份。除上述提到之外，若此次并非该车企销量首次超过 4500，则该车企应于上述连续三年之后的第一年开始，履行（b）(1)（A）条款中规定的较大型车企车队平均排放目标。

3. 若某一车企先前连续三年在加州新产 PC、LDT、MDV、重型车辆（HDV）及重型发动机的平均销售量低于 4500（辆/台），这类车企应被视为小型车企并应于下一年款开始履行小型车企的目标。

（D）ZEV 的处理方式. 根据 1962. 1 和 1962. 2 条款中规定，应履行 PC 和 LDT 车型（0 – 3750 lbs. LVW）ZEV 目标的 LDT（> 3750 lbs. LVW）型 ZEV，在计算车队平均 NMOG + NOx 值时，应计入 LDT1。

（2）PC、LDT 及 MDPV 的 LEV III 渐进式目标. 2015 – 2016 年款内，1961（a）(1) 条款规定的适用于 PC、LDT 及 MDPV 的 LEV II SULEV 排放标准只适用于沿用前一年款的"结转"排放测试数据通过 SULEV 排放标准认证的的 PC、LDT 及 MDPV 车辆，以上排放测试数据的测定基于 1982 年 11 月 16 日通过并于 1988 年 1 月 21 日修订的《美国环境保护局 OMS 第 17 号咨询通告》，特此将上述文件列入本规则内作参照。自 2017 年款起，1961（a）(1) 条款规定的 PC、LDT 及 MDPV 车型 LEV II SULEV 排放标准只适用于根据《加州乘用车、轻型卡车和中型车辆 2009 – 2017 年款零排放车和混合动力电动车尾气排放标准和测试程序》中规定获得了 PZEV 配额的 PC、LDT 及 MDPV 车辆。除小型车企之外，所有车企须保证其 2020 年及后续年款中所产全部 PC、LDT 及 MDPV 车辆遵守（a）(1) 条款中所规定的 LEV III 排放标准；小型车企须保证其 2022 年及后续年款中所产全部 PC、LDT 及 MDPV 车队遵守（a）(1) 条款中所规定的 LEV III 排放标准。

（3）MDV（不包括MDPV）LEV III 渐进式目标.

（A） 除小型车企外的 MDV 车企，须保证其 MDV 车队实现以下渐进式目标：

年款	遵守 1961.2 （a）（1） 条款的车辆（%）				遵守 1956.8 （c）或（h）条款的车辆（%）
	LEV II LEV、LEV III LEV395 或 LEV 630	LEV II ULEV、LEV III ULEV340 或 ULEV570	LEV III ULEV250 或 ULEV400	LEV III SULEV170 或 SULEV230	ULEV
2015	40	60	0	0	100
2016	20	60	20	0	100
2017	10	50	40	0	100
2018	0	40	50	10	100
2019	0	30	40	30	100
2020	0	20	30	50	100
2021	0	10	20	70	100
2022 +	0	0	10	90	100

（B）小型车企目标。2015 – 2017 年款内，小型车企须保证其生产并在加州交付销售的全部 MDV 车辆或发动机符合 MDV LEV II LEV 排放标准或 LEV III LEV395 或 LEV III LEV630 排放标准；2018 – 2021 年款内，小型车企须保证其在加州生产并交付销售的全部 MDV 车辆或发动机符合 MDV LEV II ULEV 排放标准或 LEV III ULEV340 或 LEV III ULEV570 排放标准；2022 及后续年款内，小型车企须保证其生产并在加州交付销售的全部 MDV 车辆或发动机符合 MDV LEV III ULEV250 或 LEV III ULEV400 排放标准。符合这些 MDV 排放标准的发动机不计入平均排放量。

（C）LEV III MDV 车型可选渐进式目标。2016 及后续年款内，生产并在加州交付销售四个或少于四个 MDV 测试组的车企可执行以下 LEV III MDV 车型可选渐进式目标。

1. 生产并在加州交付销售四个 MDV 测试组的车企可执行以下 LEV III MDV 车型可选渐进式目标。

年款	遵守 1961.2（a)(1) 条款的测试组数量				遵守 1956.8（c）或（h）条款的车辆（%）
	LEV II LEV、LEV III LEV395 或 LEV630	LEV II ULEV、LEV III ULEV340 或 ULEV570	LEV III ULEV250 或 ULEV400	LEV III SULEV170 或 SULEV230	ULEV
2016－2017	1	2	1	0	100
2018	0	2	2	0	100
2019	0	1	2	1	100
2020	0	1	1	2	100
2021	0	0	1	3	100
2022＋	0	0	0	4	100

2. 在加州生产并交付销售三个 MDV 测试组的车企可执行以下 LEV III MDV 车型可选渐进式目标。

年款	遵守 1961.2（a)(1) 条款的测试组数量				遵守 1956.8（c）或（h）条款的车辆（%）
	LEV II LEV、LEV III LEV395 或 LEV 630	LEV II ULEV、LEV III ULEV340 或 ULEV570	LEV III ULEV250 或 ULEV400	LEV III SULEV170 或 SULEV230	ULEV
2016	1	2	0	0	100
2017	0	2	1	0	100
2018	0	1	2	0	100
2019－2020	0	1	1	1	100
2021	0	0	1	2	100
2022＋	0	0	0	3	100

3. 在加州生产并交付销售两个 MDV 测试组的车企可执行以下 LEV III MDV 车型可选渐进式目标。

年款	遵守 1961.2（a）(1) 条款的测试组数量				遵守 1956.8（c）或（h）条款的车辆（%）
	LEV II LEV、LEV III LEV395 或 LEV 630	LEV II ULEV、LEV III ULEV340 或 ULEV570	LEV III ULEV250 或 ULEV400	LEV III SULEV170 或 SULEV230	ULEV
2016	1	1	0	0	100
2017 – 2019	0	1	1	0	100
2020 – 2021	0	0	1	1	100
2022 +	0	0	0	2	100

4. 在加州生产并交付销售一个 MDV 测试组的车企可执行以下 LEV III MDV 车型可选渐进式目标。

年款	遵守 1961.2（a）(1) 条款的测试组数量				遵守 1956.8（c）或（h）条款的车辆（%）
	LEV II LEV、LEV III LEV395 或 LEV 630	LEV II ULEV、LEV III ULEV340 或 ULEV570	LEV III ULEV250 或 ULEV400	LEV III SULEV170 或 SULEV230	ULEV
2016 – 2018	0	1	0	0	100
2019 – 2021	0	0	1	0	100
2022 +	0	0	0	1	100

（D）确定车企的 MDV 车队。某车企的"MDV 车队"即为该车企生产并在加州交付销售的加州认证 MDV 车辆的总数量。根据其所占比例，可以计算出该车企在加州交付销售的加州认证 MDV 车辆总产量。选择履行 1956.8（c）或（h）条款中可选中型发动机标准的车企，不得将此类发动机计入本节中其加州认证 MDV 车辆总产量之内。

（E）选择履行 1956.8（c）或（h）CCR 章节第 13 主题中可选中型发动机标准的车企，所有 MDV 车辆（包括小型车企所产车辆在内），都须遵守重型柴油机或奥托循环发动机适用的平均排放规定。《加州 2004 及后续年款重型奥托循环发动机的尾气排放标准和测试程序》或《加州 2004 及后续年款重型柴油发动机尾气排放标准和测试程序》对此做出规定，特此将上述文件列入 1956.8（b）节或（d）节中作参照。

（4）*SFTP 渐进式目标*。

（A）*PC、LDT 及 MDPV 渐进式目标*。若某一测试组符合 150 000 英里耐久性 LEV III FTP 排放标准，则该测试组也须同时满足 150 000 英里耐久性 SFTP 排放标准。

为满足 NMOG + NO_x 和 CO 的 SFTP 渐进式排放标准，车企有两种可选途径。

1. 途径 1：自 2015 年款起，若某一车企所生产的 PC、LDT 及 MDPV 车辆符合 150 000 英里耐久性的 LEV III FTP 排放标准，则该车企须保证此类车辆同时符合（a）(7)（A）1 条款中规定的 NMOG + NO_x 和 CO 的 SFTP 排放标准。

2. 途径 2：2015 及后续年款内，车企须保证其售出的 PC、LDT 及 MDPV 销售量加权的车队平均 NMOG + NOx 综合排放值和每个测试组的 CO 综合排放值不超过（a）(7)（A）2 条款中规定的当年款实际适用综合排放标准。

自 2017 年款起，车企须保证其所生产的、符合 150 000 英里的 LEV III FTP PM 排放标准的 PC、LDT 及 MDPV 车辆符合（a）(7)（B）条款中所规定的 SFTP PM 排放标准。

（B）*MDV 车企渐进式目标*。NMOG + NO_x 和 CO 排放标准渐进式目标从 2016 年款开始实施。对于 8501 – 10 000 lbs. GVWR 的 MDV 车辆，每年款符合 1961.2 节中 SFTP 排放标准的 MDV 车辆比例应大于或等于符合 FTP ULEV250、ULEV200、SULEV170 及 SULEV150 排放标准的 MDV 车辆所占总比例；此类车辆中，符合 SFTP SULEV 排放标准的 MDV 车辆比例应大于或等于符合 FTP SULEV170 及 SULEV150 放标准的 MDV 车辆所占总比例。对于 10 001 – 14 000 lbs. GVWRMDV 车辆，每年款符合 1961.2 节中 SFTP 排放标准的 MDV 车辆比例应大于或等于符合 FTP ULEV400、ULEV270、SULEV230 及 SULEV200 排放标准的 MDV 车辆所占总比例；此类车辆中，符合 SFTP SULEV 排放标准的 MDV 车辆比例应大于或等于符合 FTP SULEV230 及 SULEV200 放标准的 MDV 车辆所占总比例。

此外，2017 及后续年款内符合 150 000 英里耐久性的 LEV III FTP PM 排放标准的 MDV 车辆须符合（a）(7)（D）条款中所规定的 SFTP 排放标准。

（C）确定某车企的 MDV 车队规模。在 2016 及后续年款内，每个车企的 "MDV 车队" 指的是该车企生产并在加州交付销售的加州认证 MDV 车辆的总数量（MDPV 除外）。2016 及后续年款内，选择履行 1956.8（c）或（h）条款中可选中型发动机排放标准的车企，不得将此类发动机计入本小节中其加州认证 MDV 车辆总产量之内。

（c） NMOG + NO$_x$积分与借记计算

（1） PC、LDT 及 MDPVNMOG + NO$_x$积分与借记的计算。

（A） 2015 及后续年款内，某车企计算其积分或借记的方式如下：

$[$（车队平均 NMOG + NO$_x$排放目标)—(该车企车队实际 NMOG + NO$_x$平均排放量)$]×$（生产并在加州交付销售的车辆总量(含 ZEV 与 HEV 在内)）

（B） 2015 及后续年款内，若某车企的车队平均 NMOG + NO$_x$值低于本年款内规定的车队平均 NMOG + NO$_x$排放目标，则该企业会获得相应积分，以 g/mi NMOG + NO$_x$（每英里 NMOG + NO$_x$排放克数）为计算单位。但如果 2015 及后续年款内某车企的平均 NMOG + NO$_x$排放量超过了该年款的排放目标，就会产生借记，依旧以 g/mi NMOG + NO$_x$为单位，等于依据上述公式计算出来的负积分值。某一车企所生产的所有车型，包括所有 PC、0 – 3750 lbs. LVW 的 LDT 或 3 751lbs. LVW – 8 500 lbs. GVW 的 LDT 及所有 MDPV，其 g/mi NMOG + NO$_x$获得的积分或借记都会综合计算。综合计算结果即为该车企该年款内获得的 g/mi NMOG + NO$_x$积分或借记。

（2） MDV（MDPV 除外）车辆等值 NMOG + NO$_x$积分计算。

（A） 2016 及后续年款内，若某车企生产并在加州交付销售 MDV（MDPV 除外）的数量超过符合（a）（1）条款中所规定尾气排放标准的 LEV III 车辆目标，则该车企将获得车辆等值积分（VEC）。其计算方式如下公式所示（其中，"生产"指的是"生产并在加州交付销售"）：

（1.00）× |[（除 HEV 外的 LEV395 及 LEV630 产量）+
（LEV395 型 HEV 产量× LEV395 车型的 HEV VEC 系数）+
（LEV630 型 HEV 产量× LEV630 车型的 HEV VEC 系数）–
（要求生产 LEV395 及 LEV630 的数量）| +

（1.14）× |[（除 HEV 外的 ULEV340 及 ULEV570 产量）+
（ULEV340 型 HEV 产量× ULEV340 车型的 HEV VEC 系数）+
（ULEV570 型 HEV 产量× ULEV570 车型的 HEV VEC 系数）–
（要求生产 ULEV340 及 U LEV570 的数量）| +

（1.37）× |[（除 HEV 外的 ULEV250 及 ULEV400 产量）+
（ULEV250 型 HEV 产量× ULEV250 车型的 HEV VEC 系数）+
（ULEV400 型 HEV 产量× ULEV400 车型的 HEV VEC 系数）–
（要求生产 ULEV250 及 U LEV400 的数量）| +

（1.49）× |[（除 HEV 外的 ULEV200 及 ULEV270 产量）+
（ULEV200 型 HEV 产量× ULEV200 车型的 HEV VEC 系数）+

(ULEV270 型 HEV 产量 × ULEV270 车型的 HEV VEC 系数) −

(要求生产 ULEV200 及 ULEV270 的数量)} +

$(1.57) \times \{[$(除 HEV 外的 SULEV170 及 SULEV230 产量)+

(SULEV170 型 HEV 产量 × SULEV170 车型的 HEV VEC 系数)+

(SULEV230 型 HEV 产量 × SULEV230 车型的 HEV VEC 系数)−

(要求生产 SULEV170 及 SULEV230 的数量)} +

$(1.62) \times \{[$(除 HEV 外的 SULEV150 及 SULEV200 产量)+

(SULEV150 型 HEV 产量 × SULEV150 车型的 HEV VEC 系数)+

(SULEV200 型 HEV 产量 × SULEV200 车型的 HEV VEC 系数)−

(要求生产 SULEV150 及 SULEV200 的数量)} +

$[(2.00) \times ($按照 MDV 生产并得到认证的 ZEV 的数量$)]$。

（B）MDV HEV VEC 系数。MDV HEV VEC 系数计算方法如下：

LEV395：$1 + \left[\dfrac{(LEV395\ \text{排放标准} - ULEV340\ \text{排放标准}) \times \text{零排放}\ VMT\ \text{配额}}{LEV395\ \text{排放标准}} \right]$；

ULEV340：$1 + \left[\dfrac{(ULEV340\ \text{排放标准} - ULEV250\ \text{排放标准}) \times \text{零排放}\ VMT\ \text{配额}}{ULEV340\ \text{排放标准}} \right]$；

ULEV250：$1 + \left[\dfrac{(ULEV250\ \text{排放标准} - ULEV200\ \text{排放标准}) \times \text{零排放}\ VMT\ \text{配额}}{ULEV250\ \text{排放标准}} \right]$；

ULEV200：$1 + \left[\dfrac{(ULEV200\ \text{排放标准} - SULEV170\ \text{排放标准}) \times \text{零排放}\ VMT\ \text{配额}}{ULEV200\ \text{排放标准}} \right]$；

SULEV170：$1 + \left[\dfrac{(SULEV170\ \text{排放标准} - SULEV150\ \text{排放标准}) \times \text{零排放}\ VMT\ \text{配额}}{SULEV170\ \text{排放标准}} \right]$；

SULEV150：$1 + \left[\dfrac{(SULEV150\ \text{排放标准} - ZEV\ \text{排放标准}) \times \text{零排放}\ VMT\ \text{配额}}{SULEV150\ \text{排放标准}} \right]$；

LEV630：$1 + \left[\dfrac{(LEV630\ \text{排放标准} - ULEV570\ \text{排放标准}) \times \text{零排放}\ VMT\ \text{配额}}{LEV630\ \text{排放标准}} \right]$；

ULEV570：$1 + \left[\dfrac{(ULEV570\ \text{排放标准} - ULEV400\ \text{排放标准}) \times \text{零排放}\ VMT\ \text{配额}}{ULEV570\ \text{排放标准}} \right]$；

ULEV400：$1 + \left[\dfrac{(ULEV400\ \text{排放标准} - ULEV270\ \text{排放标准}) \times \text{零排放}\ VMT\ \text{配额}}{ULEV400\ \text{排放标准}} \right]$；

ULEV270：$1 + \left[\dfrac{(ULEV270\ \text{排放标准} - ULEV230\ \text{排放标准}) \times \text{零排放}\ VMT\ \text{配额}}{ULEV270\ \text{排放标准}} \right]$；

SULEV230：$1 + \left[\dfrac{(SULEV230\ \text{排放标准} - SULEV200\ \text{排放标准}) \times \text{零排放}\ VMT\ \text{配额}}{SULEV230\ \text{排放标准}} \right]$；

SULEV200：$1 + \left[\dfrac{(SULEV200\ \text{排放标准} - ZEV\ \text{排放标准}) \times \text{零排放}\ VMT\ \text{配额}}{SULEV200\ \text{排放标准}} \right]$。

此处 HEV 的"零排放 VMT 配额"根据《加州乘用车、轻型卡车和中型车辆 2009－2017 年款零排放车和混合动力电动车尾气排放标准和测试程序》与《加州乘用车、轻型卡车和中型车辆 2018 及后续年款零排放车和混合动力电动车尾气排放标准和测试程序》第 C 节中的适用性规定为准，特此将上述文件列入 1962.2 节内作参照；此外，在本（c）（2）（B）小节内，上述公式中最大许可零排放 VMT 配额可为 1.0。

（**C**）若某车企未能生产并在加州交付销售等量的符合 LEV Ⅲ 排放标准的 MDV 车辆，则该车企会产生车辆等值借记（VED），数量为依据（c）（2）（A）小节中所列公式计算出来的负积分值。

（**D**）只有认证为 MDV 且未计入 ZEV 目标的 ZEV 车辆可以被纳入 VEC 的计算。

（3）补偿借记的程序。

（**A**）某车企要补偿借记，可以通过获取与 g/mi NMOG＋NO$_x$ 借记或 VED 等量的 g/mi NMOG＋NO$_x$ 排放积分或是 VEC，或者向执行官员提交等量的 g/mi NMOG＋NO$_x$ 排放积分或是 VEC，这些积分可以是之前年款所获积分，也可以由该车企向其他车企购买所得。车企须在三个年款期间偿清 PC、LDT、MDPV 车型的 NMOG＋NO$_x$ 借记以及 MDV 车型的 VED。若在规定的时间段内未偿清，则该车企将会受到依据《健康和安全法》第 43211 节中的民事处罚，该处罚适用于那些所售新车辆未达规定的适用排放标准的车企。当所欠积分截至规定时间依然未能偿清，则有正当理由对该车企进行相应的处罚。若某一车企选择执行（b）（1）（A）1.a 小节中的"途径 2"，则在计算所欠积分时须分别计算加州、哥伦比亚特区和（b）（1）（A）1.a 小节中受车队平均温室气体排放要求约束的各州数据，并依据《健康和安全法》第 43211 节中的民事处罚分别接受处罚。该车企须依据（c）（1）或（c）（2）小节中的公式分别计算加州、哥伦比亚特区和其他各州所欠积分，除此之外，哥伦比亚特区及其他各州"生产并在加州交付销售的车辆的总量（含 ZEV 与 HEV 在内）"一项应分别计算。

根据《健康和安全法》第 43211 节，不符合排放标准的 PC、LDT 及 MDPV 的数量，为（在最初产生借记的相应年款内所有 PC、0－3 750 lbs. LVW 的 LDT、3 751 lbs. LVW－8 500 lbs. GVW 的 LDT 及所有 MDPV 的车队平均 g/mi NMOG＋NO$_x$ 排放目标）/（g/mi NMOG＋NO$_x$ 排放所欠积分总和）之商；不符合排放标准的 MDV 数量等于 VED 之值。

（**B**）在任一年款获得的排放积分在其后续 5 个年款期间始终全部有效；但从积分获得后的第 6 个年款开始，积分作废。

（4）NMOG 积分/ 借记转换为 NMOG + NO$_x$ 积分/ 借记。在 2015 年款开始之前未使用的积分或未偿清的所欠积分都会在 2015 年款之初转换为 NMOG + NO$_x$ 积分，转化后的积分值为原积分值乘以 3.0。其中，积分和借记依照 1961（c）（3）小节中所规定的公式计算。

（d）测试程序。本节中排放标准的认证目标与测试程序在《加州乘用车、轻型卡车和中型车辆 2015 及后续年款的标准污染物排放标准和测试程序及 2017 及后续年款的加州温室气体排放标准和测试程序》（2012 年 12 月 6 日修订）和《加州非甲烷有机气体测试标准》（2012 年 12 月 6 日修订）列出，特此将上述文件列于此作参照。对于混合动力车和车载燃料加热器，本节中排放标准的认证目标与测试程序在《加州乘用车、轻型卡车和中型车辆 2009 - 2017 年款零排放车和混合动力电动车尾气排放标准和测试程序》（特此将上述文件列于 1962.1 节内作参照）与《加州乘用车、轻型卡车和中型车辆 2018 及后续年款零排放车和混合动力电动车尾气排放标准和测试程序》（特此将上述文件列于 1962.2 节内作参照）。

（e）缩写。在 1961.2 中使用下述缩略语：

"ALVW" 指校正的车辆满载重量。

"ASTM" 指美国测试与材料协会。

"CO" 指一氧化碳。

"FTP" 指联邦测试程序。

"g/mi" 指克/英里。

"GVW" 指车辆总重量。

"GVWR" 指车辆额定总重量。

"HEV" 指混合动力电动车。

"LDT" 指轻型卡车。

"LDT1" 指满载重量为 0 - 3750 磅的轻型卡车。

"LDT2" 指满足 LEV II 排放标准的满载重量为 3 751 - 8 500 磅的轻型卡车，或者满足 LEV I 排放标准的满载重量为 3 751 - 5 750 磅的轻型卡车。

"LEV" 指低排放车辆。

"LPG" 指液化石油气。

"LVW" 指车辆满载重量。

"MDPV" 指中型乘用车。

"MDV" 指中型车辆。

"NMHC" 指非甲烷碳氢化合物。

"mg/mi" 指毫克/英里。

"非甲烷有机气体"或"NMOG"是指被氧化和未被氧化的碳氢化合物的总质量。

"NO_x"指氮氧化物。

"PC"指乘用车。

"SULEV"指特超低排放型车辆。

"ULEV"指超低排放型车辆。

"VEC"指汽车等值积分。

"VED"指汽车等值借分。

"VMT"指车辆行驶里程。

"ZEV"指零排放车。

（f）可分割性。本节中的每条规定都是可分割的，如本节的某一条规定被认定为无效，本节及其余的所有规定仍然继续完全有效。

注意：文献引用：《健康和安全法》第 39500、39600、39601、43013、43018、43101、43104/43105 和 43106 节。文献参考：《健康和安全法》第 39002、39003、39667、43000、43009.5、43013、43018、43100、43101、43101.5、43102、43104、43105、43106、43204 和 43205 节。

§ 1961.3　2017 及后续年款乘用车、轻型卡车和中型车辆温室气体排放标准和测试程序

简介：本 1961.3 款规定了 2017 及后续年款的乘用车（PC）、轻型卡车（LDT）和中型乘用车（MDPV）的温室气体排放水平。3 751 磅（lbs）满载重量（LVW）–8 500 lbs. 车辆总重量（GVW）符合 1961（a）（1）款中所述路径 1 下 LEV II 氮氧化物（NOx）标准的 LDT，免除遵守本条所述的温室气体排放要求，但是，PC、0–3750 lbs. LVW 的 LDT 以及 MDPV 不能免除。

紧急用车可以排除在此温室气体排放要求之外。车企必须在适用年款开始之前，以书面方式告知执行官员车企的选择，否则必须遵守本 1961.3 条款。

（a）温室气体排放目标

（1）PC、LDT 和 MDPV 车队平均二氧化碳（CO₂）目标。根据本（a）（1）款的规定，每个年款适用的 CO_2 车队平均排放标准是按销售量加权的平均的 CO_2 排放目标值。每个年款的按销售量加权的平均的 CO_2 排放值不能超过按销售量加权的平均的 CO_2 排放目标值。

（A）PC 的车队平均 CO₂ 目标值。生产并在加州交付销售的每个年款的 PC CO_2 总排放目标的平均值应当符合以下标准：

1. 对于足迹面积小于等于 41 平方英尺的 PC，克/英里（g/mi）CO_2 目标值对照下表：

年款	CO₂目标值（g/mi）
2017	195.0
2018	185.0
2019	175.0
2020	166.0
2021	157.0
2022	150.0
2023	143.0
2025	137.0
2025 及以后	131.0

2. 对于足迹面积大于 56 平方英尺的 PC，g/mi CO_2 目标值可参照下表：

年款	CO₂目标值（g/mi）
2017	263.0
2018	250.0
2019	238.0
2020	226.0
2021	215.0
2022	205.0
2023	196.0
2025	188.0
2025 及其后	179.0

3. 对于足迹面积大于 41 平方英尺但小于或等于 56 平方英尺的 PC，g/mi CO_2目标值应当按照下列公式计算，并就近保留至 0.1 g/mi：

$$目标克\ CO_2/mi = [a \times f] + b$$

其中：

f 为车辆足迹面积，以及系数 a 和 b 从以下表格选择。

年款	a	b
2017	4.53	8.9
2018	4.35	6.5
2019	4.17	4.2
2020	4.01	1.9
2021	3.84	−0.4
2022	3.69	−1.1
2023	3.54	−1.8
2025	3.4	−2.5
2025 及其后	3.26	−3.2

（B）LDT 和 MDPV 车队平均 CO_2 目标值。生产并在加州交付销售的每个年款的 LDT 和 MDPV 的 CO_2 车队平均目标值应当按照以下方式计算：

1. 对于足迹面积小于或等于 41 平方英尺的 LDT 和 MDPV，g/mi CO_2目标值可参照下表：

年款	CO_2目标值（g/mi）
2017	238.0
2018	227.0
2019	220.0
2020	212.0
2021	195.0
2022	186.0
2023	176.0
2025	168.0
2025 及其后	159.0

2. 对于足迹面积大于 41 平方英尺且小于或等于下表中对应年款的最大足迹面积的 LDT 和 MDPV，g/miCO_2目标值应当按照下列公式计算，并就近保留至 0.1g/mi：

$$目标克 CO_2／英里 = [a \times f] + b$$

其中：

f 为车辆足迹面积，以及系数 a 和 b 从以下表格对应的年款选择。

年款	最大足迹面积	a	b
2017	50.7	4.87	38.3
2018	60.2	4.76	31.6
2019	66.4	4.68	27.7
2020	68.3	4.57	24.6
2021	73.5	4.28	19.8
2022	74.0	4.09	17.8
2023	74.0	3.91	16.0
2025	74.0	3.74	14.2
2025 及其后	74.0	3.58	12.5

3. 对于某一年款足迹面积大于下表中的最小足迹面积且小于或等于下表中的最大足迹面积的 LDT 和 MDPV，其 g/mi CO_2 目标值应当按照下列公式计算，就近取整至 0.1 g/mi。

$$目标克 CO_2/mi = [a \times f] + b$$

其中：

f 为车辆足迹面积，以及系数 a 和 b 从以下表格对应的年款选择。

年款	最小足迹面积	最大足迹面积	a	b
2017	50.7	66.0	4.04	80.5
2018	60.2	66.0	4.04	75.0

4. 对于足迹面积大于下表中每个年款的最小值的 PC 和 MDPV，其 g/mi CO_2 目标值应当从以下表格对应的年款选择：

年款	最小足迹面积	CO_2目标值（g/mi）
2017	66.0	347.0
2018	66.0	342.0
2019	66.4	339.0
2020	68.3	337.0
2021	73.5	335.0
2022	74.0	321.0
2023	74.0	306.0
2025	74.0	291.0
2025 及其后	74.0	277.0

（C）适用于车企的 CO_2 车队平均标准的计算。 每个年款中，每个车企应当遵守 PC 和 LDT 以及 MDPV 适用的 CO_2 车队平均标准，每个年款的标准计算如下。每个年款中，车企必须按照（a）（A）款中所述的 CO_2 目标值分别为其 PC 车队和 LDT 与 MDPV 联合车队计算 CO_2 平均值。这些计算出的 CO_2 值就是该年款 PC 和 LDT 以及 MDPV 适用的、车企的 CO_2 平均标准。

1. 每个特定型号和足迹面积组合的 CO_2 目标值应当按照（a）（1）（A）或（a）（1）（B）计算。

2. 每个特定型号和足迹面积组合的 CO_2 目标值应当乘以适用年款的该型号/足迹面积组合的总产量。

3. 计算出的结果加总，总和被该年款适用的 PC 总产量或 LDT 与 MDPV 的联合总产量整除，结果就近取整，单位为 g/mi。该结果若适用，就是车企的 PC 车队或者 LDT 与 MDPV 联合车队的 CO_2 平均标准。

（2） PC、LDT 和 MDPV 氧化亚氮（N_2O）和甲烷（CH_4）排放标准。每个车企的所有的 PC、LDT 和 MDPV 必须遵守（a）（2）（A）、（a）（2）（B）或（a）（2）（C）款所述的 N_2O 和 CH_4 标准。除非执行官员提前批准，否则车企不能在同一年款中同时使用（a）（2）（A）和（a）（2）（B）条款。比如，车企不能对 PC 车队使用（a）（2）（A）的条款，同时对同一年款的 LDT 和 MDPV 车队使用（a）（2）（B）的条款。车企可以在同一年款中同时使用（a）（2）（A）和（a）（2）（C）款。比如，车企可符合（a）（2）（A）1 款中的 N_2O 标准以及（a）（2）（C）款中的可选 CH_4 标准。

（A） 每个测试小组适用的标准。

1. 在车辆的全使用寿命中，N_2O 排放不能超过按照 FTP（40 CFR，Part 86，Subpart B）测试（经《加州乘用车、轻型卡车和中型车辆 2015 及后续年款尾气排放标准和测试程序以及 2017 及后续年款温室气体排放标准和测试程序》修订）的 0.010 g/mi。按照（a）（2）（C）款的规定，车企还可以选择 N_2O 的可选标准。

2. 在车辆的全使用寿命中，CH_4 排放不能超过按照 FTP（40 CFR，Part 86，Subpart B）测试（经《加州乘用车、轻型卡车和中型车辆 2015 及后续年款尾气排放标准和测试程序以及 2017 及后续年款温室气体排放标准和测试程序》修订）的 0.030 g/mi。按照（a）（2）（C）款的规定，车企还可以选择 CH_4 的可选标准。

（B） 将 N_2O 和 CH_4 纳入车队平均计划中。车企可以选择不达到（a）（2）（A）款所述的排放标准。这样选择的车企必须在车辆的全使用寿命中、对每个特定型号和足迹面积的组合条件下、同时测量 FTP 测试循环和公路燃料经济测试循环中的 N_2O 和 CH_4 排放，并将测量出的 N_2O 排放值乘以 298，将 CH_4 排放值乘以 25，并按照（a）（2）（A）（D）款的规定，将 N_2O 和 CH_4 值包括在 PC、LDT 和 MDPV 车队的平均值计算中。

（C） 可选 N_2O 和/或 CH_4 标准的选择性使用。车企可以选择适用于测试小组的可选 N_2O 和/或 CH_4 标准。比如，车企可以选择达到（a）（2）（A）1 款中的 N_2O 标准，以及可选 CH_4 标准，来代替（a）（2）（A）2 款中的标准。每个污染物的可选标准都没有（a）（2）（A）款中适用的排放标准严格。可选 N_2O 和 CH_4 标准适用于在车辆的全使用寿命、按照 FTP（40 CFR，Part 86，Subpart B）计算的（经《加州乘用车、轻型卡车和中型车辆 2015 及后续年款尾气排放标准和测试程序以及 2017 及后续年款温室气体排放标准和测试程序》修订）排放量，并可作为测试小组适用的核证和在用排放标准。使用可选 N_2O 和/或 CH_4 标准的车企必须根据（a）（2）（D）款为每个测试小组计算排放借记，或与可选标准合并计算排放借记。根据（b）（1）（B）款要求，借记必须包括在一个年款的总积分或借记的计算中。灵活燃油车辆

（或者认证过多种燃油的其他车辆）在进行所有可用燃油类型的测试时也必须符合这些可选标准。

（D）CO₂当量借记。测试小组中，根据（a）（2）（C）款使用可选 N_2O 和/或 CH_4 标准的 CO_2 当量借记应按照下面的公式计算，就近取整，单位为 g/mi：

$$借记 = 全球变暖潜能值（GWP）×（产量）×（可选标准 - 标准）$$

其中：

借记 = 测试小组中，使用可选 N_2O 和/或 CH_4 标准的 N_2O 或 CH_4 CO_2 当量借记；

GWP = 25（计算 CH_4 借记时）；298（计算 N_2O 借记时）；

产量 = 测试小组中生产并在加州交付销售的车辆数量；

可选标准 = 生产商按照（a）（2）（C）款选择的可选标准（N_2O 或 CH_4）；以及

标准 =（a）（2）（A）款所述 N_2O 或 CH_4 排放标准。

（3）美国销售量受限的车企的可选车队平均标准。符合本（a）（3）款标准的车企可以要求执行官员设立另外的、替代（a）（1）款中的 CO_2 车队平均标准。

（A）申请可选标准的资格。本（a）（3）款规定的资格应当基于 PC、LDT和 MDPV 的总销售量。本（a）（3）款所说的"销售量"和"已售"指的是在美国的各州和领土上生产并交付销售（或已经售出）的车辆。为确定资格，相关公司的销售量应当根据 1900 章进行汇总。为获得申请本（a）（3）款所规定的可选标准的资格，车企最近连续三个年款的平均销售量必须小于 5 000 辆。如果车企最近连续三个年款的平均销售量超过 4 999，则车企将不具备免除资格，且必须符合以下适用的排放标准。

1. 如果一家车企连续三个年款的平均销售量超过 4 999 辆，同时，如果销售量上涨是由企业购并、合并或被另外一家车企购买导致，则该车企必须从连续三个年款后的第一个年款开始遵守（a）（1）和（a）（2）款中所述适用的排放标准。

2. 如果一家车企连续三年的平均销量超过 4 999 辆但低于 50 000 辆，同时，如果销售量上涨仅是由车企的产能扩张，而不是由企业购并、合并或被另外一家车企购买导致，则车企必须从连续三个年款后的第二个年款开始遵守（a）（1）和（a）（2）款中所述适用的排放标准。

（B）对新进入美国市场的车企的要求。新进入车企是指以前在美国没有车辆销售记录并且没有达到 40 CFR §86.1818 - 12 中规定的或者没有达到 1961.1 节中的温室气体标准的车企。除了符合（a）（3）（A）款中的资格要求，新进入车企还应当符合以下要求：

1. 除了（a）（3）（D）款中要求提交的信息，新进入车企还必须提供相关文件，清楚地表明公司在可选标准适用年份间进入美国市场的意愿。表明这种意愿的方式，包括提供证明建立美国销售网络的文件、证明公司正在为达到其他美国目标（比如安全标准）而努力的文件，或者其他合理的、证明该意愿的、让执行官员满意的信息。

2. 新进入车企在美国市场的最初两个年款的销售量必须在 5 000 辆以下，并且在进入美国市场的最初五年中，任何连续三年的平均销售量应当低于 5 000 辆。超出这些限制出售的车辆不会被授予许可证，且车企将为每辆没有许可证的车受到惩罚。此外，超出这些限制，车企会失去申请可选标准的资格，直到车企证明连续两个年款的销售量少于 5 000 辆。

3. 如果车企最近年款的销售量少于 5 000 辆，但之前年款的销售量不少于 5 000 辆，则车企有资格申请（a）（3）款所述的可选标准。但是，该车企会被看做是新进入车企，受到本（a）（3）款对新进入车企的限制，除了（a）（3）（B）（1）款中所述的证明进入美国市场意愿的要求不适用。

（C）如何申请可选车队平均标准。在信息充分的情况下，符合条件的车企可以最多连续 5 个年款申请可选标准。

1. 从 2017 年款开始，如果申请可选标准，符合条件的车企必须不迟于 2013 年 7 月 30 日上交完整的申请。

2. 从 2017 年款以后开始，如果申请可选标准，符合条件的车企必须不迟于在第一个适用可选标准的年款开始之前的 36 个月上交完整的申请。

3. 申请必须包括（a）（3）（D）款中规定的所有信息，必须由公司的首席领导签字。如果执行官员认为申请内容不完整或者不充分，车企会得到通知，并获得额外的 30 天来修改申请。

4. 车企可以选择申请（a）（3）（C）款中的可选标准，方式是将上交给 EPA 的文件（根据 40 CFR §86.1818 - 12（g）的规定，并同时参考修订过的《加州乘用车、轻型卡车和 2015 及后续年款尾气排放标准和测试程序以及 2017 及后续年款温室气体排放标准和测试程序》），以及 EPA 针对车企 2017 - 2025 年款全国温室气体计划的可选车队平均标准申请的许可，同时上交给加利福尼亚空气资源委员会（CARB）。

（D）数据和信息提交要求。申请（a）（3）款中可选标准的符合条件的车企须提交以下信息给 CARB。执行官员可以要求他/她认为合适的额外的信息。申请完成后应当送交给 CARB，地址为：加利福尼亚埃尔蒙特 9480 Telstar 大道 4 室，加利福尼亚空气资源委员会，移动污染源控制部主任，91731。

1. 车辆型号和整体信息。

a. 申请采用可选标准的年款，限制在 5 个连续年款内。

b. 每个年款的车辆型号和产量预测。

c. 每个型号的详细介绍，包括车型分类、质量、功率、足迹面积和预售价。

d. 每个型号预计的生产周期，包括新的型号推出和再设计或者更新的周期。

2. *技术评估信息*。

a. 车企每个车辆型号应用的 CO_2 减排技术，包括关于成本和 CO_2 减排效果的信息。包括改进制冷效率和减少空调系统泄漏的技术，以及任何 FTP 和 HWFET 测试操作之外的可产生效益的"非使用状态"技术。

b. 其他车企可比较型号的评估，包括由该型号产生的 CO_2 结果和空调积分。可比较的车辆应当在以下方面是相似的，但是不一定是完全相同的：车辆类型、功率、质量、功率－重量比、足迹面积、零售价和其他相关因素。对于申请自 2017 年款开始的可选标准的车企，可比较车辆的分析应当包括 2012 和 2013 年款的车辆，否则，至少应当包括最近两个年款的车辆。

c. 关于在美国市场以外的并且不在美国销售的车辆使用的减排 CO_2 技术的讨论，包括关于为什么这些车辆和/或技术没有被用来减少美国市场内的车辆的 CO_2 排放的讨论。

d. 对 CARB 在《员工报告：对加利福尼亚温室气体、标准污染物排放和挥发性排放标准和测试程序，委员会乘用车、轻型卡车和中型乘用车车载诊断系统要求，重型车辆挥发性排放标准的 LEV III 修正案的规则制定和听证会初始理由陈述》（2011 年 12 月 7 日）以及本报告附件中所提及的技术评估（这些技术可能会被用于达到温室气体排放标准），以及车企应用或计划应用这些技术的程度。对于没有计划完全应用的技术，车企应当对此作出解释。

3. *信息支持资格*。

a. 之前三个年款的美国销售量以及车企寻求可选标准的年款的预计销售量。

b. 与其他车企的所有权关系的信息，包括关于应用 40 CFR §86.1838－01（b）(3)条款和 1900 章中有关相关公司的销售加总的细节。

（E） 可选标准。在收到完整的申请后，执行官员将审核申请，确定是否可以授权可选标准。如果执行官员认为可以授权可选标准，则以下标准适用。根据本（a）(3)(E)款，"极小车企"指的是符合（a）(3)款要求的车企。

1. 从申请可选标准的年款之前的三个年款开始，每个极小车企应当在申请可选标准的年款的四个年款之前确定所有车辆型号，这些车辆型号由大型车企认定，基于车辆的款型和足迹面积，这些大型车企在申请可选标准的年款中与极小企业相当。极小车企应当向执行官员证明所选择的每个相当车款的合理性。执行官员一旦批准，就应当向极小车企提供每个款型和足迹面积的克 CO_2／英里的目标值。极小车企应当按照（a）（1）（C）款的规定、基于执行官员提供的 CO_2 目标排放值计算平均 CO_2 标准。

2. 对于 2017 及后续年款，极小车企应当符合以下任何一条件：

a. 不超过根据（a）（1）（C）计算的、基于执行官员提供的目标 CO_2 值的平均 CO_2 标准；或

b. 一旦执行官员批准，如果极小车企证明某个型号的车辆使用发动机、传动和排放控制系统，并且其足迹面积与大型车企获准在加利福尼亚销售的配置相似，那么该极小车企可不必达到 2. a 所述的标准。

（F）积分交易限制。 经执行官员批准的、受本（a）（3）款规定的可选标准限制的车企不能与其他车企交易积分。同一车企的汽车和卡车之间的积分转换是被允许的。

（4）电动车、插电式混合动力电动车和燃料电池车的温室气体排放值。

（A）电动车的计算

1. 对每个特定款型和足迹面积的组合，车企应当利用以下公式计算城市 CO_2 值：

$$城市\ CO_2\ 值 = (270\text{g}CO_2e/kWh) \times E_{EV} - 0.25 \times CO_{2\ 目标}$$

其中 E_{EV} 由使用电池电动车技术的每个测试组车辆的每个测试循环直接计算而来，单位为千瓦时每英里（SAE J1634，此处作为参考引入）

2. 对每个特定款型和足迹面积组合，车企应当利用以下公式计算公路 CO_2 值：

$$公路\ CO_2\ 值 = (270\ \text{g}CO_2e/kWh) \times E_{EV} - 0.25 \times CO_{2\ 目标}$$

其中 E_{EV} 由使用电池电动车技术的每个测试组车辆的每个测试循环直接计算而来，单位为千瓦时每英里（SAE J1634，此处作为参考引入）

（B）插电式混合动力电动车的计算。 对特定款型和足迹面积组合，车企应当利用以下公式计算城市 CO_2 值和公路 CO_2 值：

$$城市\ CO_2\ 值 = GHG_{城市}$$

以及

$$公路\ CO_2\ 值 = GHG_{公路}$$

$GHG_{城市}$ 和 $GHG_{公路}$ 根据《加州乘用车、轻型卡车和 2009 - 2017 年款零排放车和

混合动力电动车尾气排放标准和测试程序》，或者《加州乘用车、轻型卡车和中型车辆 2018 及后续年款零排放车和混合动力电动车尾气排放标准和测试程序》中第 G.12 节规定。

（C）燃料电池车的计算。对每个特定款型和足迹面积组合，车企应当利用以下公式计算城市 CO_2 值和公路 CO_2 值：

$$城市\ CO_2 = GHG_{FCV} = (9132\ gCO_2e/kg\ H_2) \times H_{FCV} - G_{上游}$$

以及

$$公路\ CO_2 = GHG_{FCV} = (9132\ gCO_2e/kg\ H_2) \times HFCV - G_{上游}$$

其中 H_{FCV} 指的是每英里消耗的氢气，单位为 kg/mi，计算依据是 SAE J2572（2008 年 10 月出版），此处作为参考引入。

（5）车队平均二氧化碳值的计算

（A）对于每个特定款型和足迹面积组合，车企按以下方式计算城市/公路 CO_2 排放总值：

$$0.55 \times 城市\ CO_2 值 + 0.45 \times 公路\ CO_2 值$$

"城市" CO_2 排放量应当按照 FTP 测试循环（40 CFR，86 部分，B 款），以及经修订的《加州乘用车、轻型卡车和中型车辆 2015 及后续年款标准污染物尾气排放标准和测试程序以及 2017 及后续年款温室气体排放标准和测试程序》测量，"公路" CO_2 排放量应当按照公路燃油经济性测试程序（HWFET；40 CFR 600 B 款）测量。

（B）每个特定款型和足迹面积组合的城市/公路 CO_2 总排放量应当乘以适用年款的款型/足迹组合的总产量。

（C）计算结果应当加总，得到的总和应当除以该年款 PC 总产量或 LDT 和 MDPV 的总产量，结果就近取整至 g/mi。这个结果应当作为车企实际的 PC 或 LDT 和 MDPV 联合的按销售量加权的车队平均 CO_2 值。

（D）对于每个年款，车企应当证明其在整个年款遵守（a）（1）款中的车队平均目标，有以下两个途径可供选择：

途径 1：符合 1961.3 规定的加州尾气排放标准的、生产并在加州交付销售的 PC、LDT 和 MDPV 总量；或者

途径 2：符合本 1961.3 款规定的加州尾气排放标准的、生产并在加州、哥伦比亚地区和所有遵守联邦《清洁空气法》（42 U.S.C. § 7507）第 177 章规定各州交付销售的 PC、LDT 和 MDPV 的总量。

1. 选择遵守途径 2 的车企应当在适用年款开始以前、以书面的形式告知执行官员，否则就要遵守途径 1。一旦车企选择途径 2，那么该选择生效，除非车企在适用年款开始以前、以书面的形式告知执行官员其将选择途径 1。

2. 如果车企某年款选择途径2，则1961.3所述的"在加州"指的是加州、哥伦比亚地区、以及所有按照联《清洁空气法》（42 U. S. C. § 7507）第177章规定通过加州温室气体排放标准的各州。

3. 选择途径2的车企必须给执行官员提供在平均范围内的，在哥伦比亚地区和第177所列各州生产并交付销售的每个款型和足迹面积的各州的值，以及适用于每个款型和足迹面积的城市 CO_2 值和高速 CO_2 排放值。

（6）空调直接排放的减排积分。车企可以在车辆使用年限内通过使用特定的用于减少空调直接排放的空调系统技术来获得空调（A/C）直接排放积分。车企获得执行官员对某型款的 A/C 直接排放积分的批准以后，只能在这种款型的车辆上使用 A/C 直接排放积分。A/C 直接排放积分获得批准的条件和要求在以下（A）至（F）款中有描述。

（A）A/C 直接排放积分的批准申请应当按照型款分类，申请应当包括：

- 车辆品牌以及
- 安装有空调系统（泄漏的积分可用的系统）的该款型车辆数量。

同一 A/C 系统中任何两个不同的配置（不包含尺寸偏差）应当分别递交申请。

（B）要获得 A/C 直接排放积分的批准，车企应当通过工程评估证明相关 A/C 系统能够减少 A/C 直接排放。该证明应当包含以下所有因素：

- 申请的 A/C 直接排放积分数，以克 CO_2 当量/英里为单位（gCO_2e/mi）；
- 根据（a）(6)(C）款计算积分数的过程；
- A/C 系统示意图；
- 系统元件的规格，要足够详细，以允许重新计算；以及
- 解释做出了何种努力，将配件和接合处数目最小化，将元件最优化，从而将泄漏最小化。

计算过程中的所有结果必须保留至少三个有效数字，最后的积分值取整到 gCO_2e/mi 的小数点后一位。

（C）A/C 直接排放积分的计算取决于制冷剂或者系统类型，在本款第1、2、3段有描述。

1. HFC - 134a 蒸汽压缩系统

使用 HFC - 134a 制冷剂的 A/C 系统，A/C 直接排放积分按照以下公式计算：

$$A/C \text{ 直接积分} = \text{直接积分基数} \times （1 - LR/Avg\ LR）$$

其中：

直接积分基数 = 12.6 gCO$_2$e/mi（PC）；

直接积分基数 = 15.6 gCO$_2$e/mi（LDT 和 MDPV）；

Avg LR（平均泄漏率）= 16.6 克/年（PC）；

Avg LR = 20.7 克/年（LDT 和 MDPV）；

LR（泄漏率）= 取 SAE 泄漏率或者最小泄漏率中较大的一个；

其中：

SAE LR = 根据国际地面车辆标准 SAE J2727（2012 年 2 月修订）计算的初始泄漏率，此处作为参考引入；

Min LR（最小泄漏率）= 8.3 克/年（PC 中使用传送带驱动压缩机的 A/C 系统）

最小泄漏率 = 10.4 克/年（LDT 和 MDPV 中使用传送带驱动压缩机的 A/C 系统）；

最小泄漏率 = 4.1 克/年（PC 中使用电动压缩机的 A/C 系统）

最小泄漏率 = 5.2 克/年（LDT 和 MDPV 中使用电动压缩机的 A/C 系统）；

注：初始泄漏率指的是新制造的 A/C 系统的制冷剂泄漏率，以克/年为单位计算。如果执行官员认为更新的 SAE J2727 或者其他方法能够产生比 2012 年 2 月版的 SAE J2727 更精确的 A/C 系统初始泄漏率估值的话，他/她可允许车企使用更新的 SAE J2727 或者其他方法。

2. 低 GWP 蒸汽压缩系统

对于使用 GWP 小于等于 150 的制冷剂的 A/C 系统，应当通过以下公式计算 A/C 直接排放积分：

$$\text{A/C 直接积分} = \text{低 GWP 积分} - \text{高泄漏罚分}$$

其中

$$\text{低 GWP 积分} = \text{最大低 GWP 积分} \times (1 - \text{GWP}/1430)$$

以及

$$\text{高泄漏罚分} = \begin{cases} \text{最大高泄漏罚分} & (\text{SAE LR} > \text{Avg LR}) \\ \text{最大高泄露罚分} \times \dfrac{\text{SAE LR} - \text{Min LR}}{\text{Avg LR} - \text{Min LR}} & (\text{Min LR} < \text{SAE LR} \leqslant \text{Avg LR}) \\ 0 & (\text{SAE LR} \leqslant \text{Min LR}) \end{cases}$$

其中：

最大低 GWP 积分 = 13.8 gCO$_2$e/mi（PC）；

最大低 GWP 积分 = 17. 2 gCO$_2$e/mi（LDT 和 MDPV）；

GWP =（a）（6）（F）款所述的 100 年内制冷剂全球变暖潜能值（GWP）；

最大高泄漏罚分 = 1. 8 gCO$_2$e/mi（PC）；

最大高泄漏罚分 = 2. 1 gCO$_2$e/mi（LDT 和 MDPV）；

平均泄漏率 = 13. 1 克/年（PC）；

平均泄漏率 = 16. 6 克/年（LDT 和 MDPV）；

其中：

SAE 泄漏率 = 根据国际地面车辆标准 SAEJ2727（2012 年 2 月修订）评估的初始
泄漏率

最小泄漏率 = 8. 3 克/年（PC）；

最小泄漏率 = 10. 4 克/年（LDT 和 MDPV）；

注：初始泄漏率指的是新制造的 A/C 系统的制冷剂泄漏率，以制冷剂克/年为单位计算。如果执行官员认为，更新的 SAE J2727 或者其他方法能够产生比 2012 年 2 月版的 SAE J2727 更精确的计算 A/C 系统初始泄漏率估值的话，他/她可允许车企使用更新的 SAE J2727 或者其他方法。

3. 其他 A/C 系统

对于使用非蒸气压缩循环技术的 A/C 系统，执行官员可以批准授予 A/C 直接排放积分。积分值应当在实施了包括可核实的实验室测试数据在内的工程评估、证明该技术能减少 A/C 的直接排放后确定，且不能超过 13. 8 gCO$_2$e/mi（PC）和 17. 2 gCO$_2$e/mi（LDT 和 MDPV）。

（D） PC、LDT 和 MDPV 的空调系统泄漏降低产生的总积分应当根据下列公式分别计算：

$$总积分(g/mi) = A/C 直接积分 × 产量$$

其中：

A/C 直接积分按照（a）（6）（C）款计算。

产量 = 生产并在加州交付销售的、安装有（a）（6）（C）款所述 A/C 直接积分值适用的空调系统的 PC、LDT 和 MDPV 总数。

（E） 根据（a）（6）（D）款计算的结果，就近取整至 g/mi，结果应当包括在车企按照（b）（1）（B）款计算的积分/借记的总数中。

（F） 下表列出的制冷剂全球变暖潜能值（GWP），或执行官员决定的其他值，应当用于本（a）（6）款的计算中。如果收到车企要求，执行官员应当根据政府间气候变化专门委员会（IPCC）的发现或者其他适用的研究结果确定不包括在本

（a）（6）（F）款中的制冷剂的值。

制冷剂	GWP
HFC – 134a	1 430
HFC – 152a	124
HFO – 1234yf	4
CO_2	1

（7）提高空调系统效率的积分。

车企可以在 PC、LDT 和/或 MDPV 使用寿命内使用特定的减少 CO_2 排放的空调系统技术来获得 CO_2 积分。车企使用产生 CO_2 积分的空调系统所产生的积分值应当按照本（a）（7）款计算。空调系统必须满足（a）（7）（E）款中规定的资格要求才能产生积分。

（A） 下表所列技术可产生 g/mi 为单位的空调效率积分，各类车辆的积分值也在下表显示：

空调技术	PC （g/mi）	LDT 和 MDPV （g/mi）
减少再加热、通过外部控制的、可变排量压缩机（比如，在乘客车厢内根据预设的气温和/或制冷要求来控制排量的压缩机）	1.5	2.2
减少再加热、通过外部控制的、固定排量或者气动可变排量压缩机（比如，通过空调系统内部或相关的条件，如水头压力、吸入压力或蒸发器出口温度，来控制排量的压缩机）	1.0	1.4
当环境温度为 75°F 或更高时，默认通过闭环回路控制空气供给来使空气再循环（通过传感器反馈来控制内部空气质量）：不同温度下，通过闭环回路控制空气供给的空调系统可以在向管理部门提交工程评估并获审批后获得积分。	1.5	2.2
当环境温度为 75°F 或更高时，默认通过开环回路控制空气供给来使空气再循环（没有传感器反馈）。不同温度下，通过开环回路控制空气供给的空调系统可以在向管理部门提交工程评估并获审批后获得积分。	1.0	1.4
限制电力浪费的鼓风电动机（如：脉宽调制电源控制器）	0.8	1.1
内部换热器（如：仪器把热量从进入蒸发机的高压、液态的制冷剂转变为蒸发机释放的低压、气态的制冷剂）	1.0	1.4

空调技术	PC (g/mi)	LDT 和 MDPV (g/mi)
经过改进的冷凝器和/或蒸发机，其零部件系统分析显示，与之前的行业标准比较，系统性能改进系数超过 10%。	1.0	1.4
油分离器。车企必须上交一份工程评估，展示系统在基本设计的基础上的改进程度，用于比较的基本部件是车企最近生产的、相同设计的、类似的或有关联的车型上应用的版本。基本部件的特点应当与新的部件相比较以显示改进。	0.5	0.7

（B） 空调效率积分在空调系统的基础上确定。每个使用（a）(7)(A) 款中一个或多个技术的空调系统，总积分值为根据（a）(7)(A) 款、适用于该空调系统的 g/mi 值的和。但是，每个空调系统的总积分值不能超过 5.0 g/mi（PC）或 7.2 g/mi（LDT 和 MDPV）。

（C） PC 和 LDT 与 MDPV 加总的空调系统产生的总效率积分应当分别根据以下公式计算：

$$总积分（g/mi）= 积分 × 产量$$

其中：

积分 = 根据适用性，以（a）(7)(B) 或（a）(7)(E) 确定的 CO_2 效率积分值

产量 = 根据适用性，生产并在加州交付销售的、安装了采用（a）(7)(B) 款所述效率积分的空调系统的 PC 或 LDT 和 MDPV 加总总量。

（D） 按照（a）(7)(C) 计算的结果，就近取整至 g/mi，应当包括在根据（b）(1)(B) 计算的车企的积分/借记总数中。

（E） 根据（a）(7)(E) 款规定，AC17 测试程序指的是 40 CFR §86.167 - 17 中提出的 AC17 空调效率检测程序，包括在《加州乘用车、轻型卡车和中型车辆 2015 及后续年款标准污染物尾气排放标准和测试程序以及 2017 及后续年款温室气体排放标准和测试程序》中，并进行了修订。

1. 车企应当在其所选择的、用于产生空调效率积分的空调系统上实施 AC17 测试程序。

2. 通过良好的工程评估，且在追求效率积分的基础上车企必须选择空调系统预期能产生最大 CO_2 排放增长的车辆配置进行测试。如果空调系统安装在 PC、LDT 和 MDPV 上，那么 PC、LDT 和 MDPV 的积分必须分别计算，但是只需测试一辆有代表性的车辆上的空调系统，前提是该车辆代表了空调系统对 CO_2 排放影响最坏的情形。

3. 对于车企选择的产生空调效率积分的每个空调系统，车企应当根据下面的要求实施 AC17 测试程序。每个空调系统应当按照下列标准进行测试：

a. 在使用能产生积分技术的空调系统的车辆上实施 AC17 测试程序

b. 在未使用能产生积分技术的空调系统的车辆上实施 AC17 测试程序。接受测试的车辆应当与（a）(7)(E)(3) a 款中的车辆相似。

c. 根据（a)(7)(E)(3) b 计算的 CO_2 排放量减去根据（a)(7)(E)(3) a 计算的 CO_2 排放量，并保留到 0.1 g/mi。如果结果小于或等于零，那么该空调系统不能产生积分。如果结果大于或等于根据（a)(7)(B) 计算的值，那么该空调系统可以产生根据（a)(7)(B) 计算的允许的最大积分值。如果结果大于零但小于（a)(7)(B) 款计算的值，那么该空调系统可以产生积分，该值由根据（a)(7)(E)(3) b 计算的 CO_2 排放量减去根据（a)(7)(E)(3) a 计算的 CO_2 排放量，并取整到 0.1 g/mi 得到。

4. 在空调系统预计产生积分的第一年款间，车企应当选择每个使用该空调系统的车辆平台中销量最高的子配置进行检测。在以后的年份，只要符合以下条件，那么该空调系统就能继续产生积分：

a. 空调系统的元件和/或控制策略没有产生可能导致效率改变的变化；

b. 车辆平台设计不变，以防引起可能导致空调系统效率变化的改变，并且

c. 在每个年款使用空调系统的平台上，车企继续测试至少一个子配置，直到每个平台上的所有子配置都经过测试。

5. 每个空调系统必须经过测试且必须符合测试要求，才能产生积分。在空调系统预期产生积分的第一年款，车企应当通过良好的工程评估选择每个使用空调系统的车辆平台中销量最高的子配置进行测试。在以后的年份，只要车企继续每年检测每个平台上的至少一个子配置，且空调系统和车辆平台没有发生大的变化，空调系统就可以继续产生积分。

（8）非使用状态积分。如果 CO_2 减排技术的效果没有在 FTP 和/或 HWFET 中得到充分体现，车企仍可以产生 CO_2 减排技术积分。这些技术应当在 FTP 和 HWFET 规定的条件之外产生可测量、可证明和可核实的实际 CO_2 减排。这些可选择的积分叫做"非使用状态"积分。用来产生减排积分的非使用状态技术被认为是与排放相关的，适用一定要求的零部件，并且必须证明在车辆使用寿命之内可有效运作。除非车企可以证明该技术不会产生使用损耗，否则车企必须在分析当中计入损耗。车企必须使用本（a)(8) 款中规定的三个选项中的一项来确定 CO_2 g/mi 的适用于某项非使用状态技术的积分。

（A）某些非使用状态技术可以产生的积分。

1. 以下表格列出了某些技术可以产生的 CO_2 积分，条件是每项技术应用于

表中车企每个年款的 PC、LDT 和 MDPV 全美总产量的最小百分比。技术定义见（e）款。

非使用状态技术	PC (g/mi)	LDT 和 MDPV (g/mi)	占全美产量的最小百分比
主动式气动布局	0.6	1.0	10
高效外部照明	1.1	1.1	10
发动机热回收	0.7 每 100W 功率	0.7 每 100W 功率	10
发动机起停（怠速停机）	2.9	4.5	10
变速器主动预热	1.8	1.8	10
发动机主动预热	1.8	1.8	10
电加热循环泵	1.0	1.5	不可提供
太阳能顶板	3.0	3.0	不可提供
热控制	≤3.0	≤4.3	不可提供

a. 按照以下表格显示的量，（e）款定义的热控制技术产生的积分可以累加：

热控制技术	积分值：PC (g/mi)	积分值：LDT 和 MDPV (g/mi)
玻璃或抛光	≤2.9	≤3.9
主动座椅通风设备	1.0	1.3
太阳能反射涂料	0.4	0.5
被动客舱通风	1.7	2.3
主动客舱通风	2.1	2.8

b. 热控制技术的最大积分限制为 3.0 g/mi（PC）、4.3 g/mi（LDT 和 MDPV）。玻璃或抛光的最大积分限制在 2.9 g/mi（PC）、3.9 g/mi（LDT 和 MDPV）。

c. 玻璃或抛光积分根据以下等式计算：

$$积分 = \left[Z \times \sum_{i=1}^{n} \frac{Ti \times Gi}{G} \right]$$

其中：

积分 = 每辆车的玻璃或抛光总积分，单位为 g/mi，PC 不能超过 3.0g/mi，LDT 和 MDPV 不能超过 4.3g/mi。

Z = 0.3（PC），0.4（LDT 和 MDPV）；

Gi = 测量的 i 窗的玻璃面积，单位为平方米，保留到小数点后一位；

G = 车辆总的玻璃面积，单位为平方米，保留到小数点后一位；

Ti = 估算的 i 窗的玻璃面积的降温值，通过以下公式计算：

$$Ti = 0.3987 \times (Tts_{base} - Tts_{new})$$

其中：

Tts_{new} = 玻璃的总阳光透光率，根据 ISO 13837：2008 "安全抛光材料—确定阳光透光率的方法" 计算（此处作为参考引入）。

Tts_{base} = 62（挡风玻璃，侧前面，侧后面，后部，以及后窗玻璃的位置），40（顶部玻璃的位置）

2. 根据（a)(8)(A) 1 的默认积分值，车企 PC、LDT 和 MDPV 平均 CO_2 排放值最大允许降幅为 10 g/mi。如果车企生产的任何一辆 PC 或 LDT 根据（a)(8)(A) 1 款表格计算的总 CO_2 积分值不超过 10 g/mi，那么总的非使用状态积分可以根据（a)(8)(D) 款计算。如果车企产品线中的任何一辆 PC、LDT 或 MDPV 根据（a)(8)(A) 1 款表格计算的总 CO_2 积分值大于 10 g/mi，那么车企 PC、LDT 和 MDPV 的 CO_2 排放降值必须根据（a)(8)(A) 2.a 计算，以便确定是否超过了 10 g/mi 的限度。

a. 按照以下公式计算 PC、LDT 和 MDPV 的 CO_2 减排值：

$$减排量 = \frac{积分 \times 1\,000\,000}{[(产量\,C \times 195\,264) + (产量\,T \times 225\,865)]}$$

其中：

积分 = PC、LDT 和 MDPV 的积分总值，单位为兆克，根据（a)(8)(D) 款计算，且仅限于（a)(8)(A) 1 款中默认的可累加的积分。

产量$_C$ = 车企生产并在美国交付销售的 PC 数量

产量$_T$ = 车企生产并在美国交付销售的 LDT 和 MDPV 数量

b. 如果根据（a)(8)(A) 2.a 款计算的积分值大于 10 g/mi，那么车企可以根据（a)(8)(A) 1 款默认的积分进行累加的总积分（单位为兆克）应当按照以下公式计算：

$$积分(兆克) = \frac{[10 \times (产量\,C \times 195\,264) + (产量\,T \times 225\,865)]}{1\,000\,000}$$

其中：

产量$_C$ = 车企生产并在美国交付销售的 PC 数量

产量$_T$ = 车企生产并在美国交付销售的 LDT 和 MDPV 数量

c. 如果根据（a）(8)（A）2. a 款计算的值不大于 10 g/mi，那么车企根据（a）(8)（A）1 款中默认的 g/mi 值累加的总积分不超过允许的范围，并且每个种类车辆的总积分可以根据（a）(8)（D）款计算。

d. 如果根据（a）(8)（A）2. a 款计算的值大于 10 g/mi，那么 PC、LDT 和 MDPV 的积分总值（根据（a）(8)（D）款计算的积分值进行累加，单位为兆克），不能超过（a）(8)（A）2. b 款中规定的值。解决途径为，只要减少导致超限的车辆种类的积分量，这样总值就不会超过（a）(8)（A）2. b 款计算的值了。

3. 如果不使用（a）(8)（A）1 款中对特定技术默认的 g/mi 值，车企还可以为该特定技术确定可替换的值。替换值必须根据（a）(8)（B）或（a）(8)（C）款中规定的方法确定。

（B）*使用 EPA 5 循环方法的技术论证*。 为证明非使用状态技术并通过 EPA 5 循环方法确定 CO_2 积分，车企需利用 40CFR 第 600 部分所述 EPA 5 循环方法确定非使用状态城市/公路碳类尾气排放效果。测试应当在有代表性的、积分对应的每个型号的车辆上进行，该车辆经过专业的工程标准选出。技术排放效果应由在非使用状态技术开启以及关闭的状态下进行的测试确定。多个非使用状态技术可以在测试车辆上论证。车企应当实施下列步骤并向执行官员提交所有的测试数据。

1. 非使用状态技术未安装和/或关闭的情况下的测试。根据 40CFR 第 600 部分 B 子部分规定的测试程序条款，利用 FTP、HWFET、US06、SC03 和低温 FTP 测试程序以及本章中 §600.113 – 08 规定的计算程序，确定碳类排放值。在非使用状态技术未安装和关闭的情况下，每个测试至少进行三遍，对每个测试的每阶段（包）的结果取平均值。根据每阶段结果的平均值计算 5 循环权重的城市/公路总排放值，其中 5 循环城市权重为 55%，5 循环公路权重为 45%。计算出的城市/公路总排放值就是该车辆的基本 5 循环碳类排放值。

2. 非使用状态技术安装和/或运行的情况下的测试。根据 40CFR 第 600 部分 B 子部分规定的测试程序条款，利用 US06、SC03 和低温 FTP 测试程序和 40CFR §600.113 – 08 款规定的计算程序，确定碳类尾气排放值。在非使用状态技术安装和运行的情况下，每个测试至少进行三遍，对每个测试的每阶段（包）的结果取平均值。根据每阶段结果的平均值计算 5 循环权重的城市/公路总碳类尾气排放值，其中 5 循环城市权重为 55%，5 循环公路权重为 45%。同时还要利用 FTP 和 HWFET 测试中根据（a）(8)（B）1 款算出的、在没有非使用状态技术的情况下的每阶段结果的平均值。最终计算出的城市/公路总排放值就是 5 循环碳类尾气排放值，该值体现了该技术的非使用状态效果，但是不包括该技术在 FTP 和 HWFET 测试中的效果。

3. 根据（a）(8)（B）2 款计算的值减去根据（a）(8)（B）1 款计算的城市/

公路排放总值，计算出的结果就是接受评估的一项或多项技术的非使用状态效果。如果最终值大于或等于根据（a）(8)（B）1 款计算的值的 3%，那么车企可以利用该值，就近取整到小数点后一位，单位为 g/mi，以便根据（a）(8)（C）款确定积分。

4. 如果按照（a）(8)（B）3 款计算的值小于根据（a）(8)（B）1 款计算的值的 2%，那么车企必须重新进行（a）(8)（B）1 和（a）(8)（B）2 款中的测试，每个测试再多进行两遍，而不用进行三次测试。（a）(8)（B）3 款中所述的接受评估的一项或多项技术的非使用状态效果应当通过（a）(8)（B）1、（a）(8)（B）和（a）(8)（B）4 款所述的所有测试进行计算。如果按照（a）(8)（B）3 款计算的值小于（a）(8)（B）1 款中计算的值的 2%，那么车企必须利用 EPA 车辆模拟工具核实其一项或多项非使用状态技术的减排潜力。如果测试出的积分值小于（a）(8)（B）1 款中计算的值的 2%，那么车企可以利用（a）(8)（B）3 款中所述的非使用状态减排效果，并通过（a）(8)（B）1、（a）(8)（B）和（a）(8)（B）4 款所述的所有测试进行计算，就近取整到小数点后一位，单位为 g/mi，并根据（a）(8)（C）款确定积分。

（C）非使用状态积分的评估和审核。

1. 前期步骤。

a. 申请（a）(8)（B）款中所述的非使用状态积分的车企必须进行该款所述的测试和/或模拟。

b. 申请（a）(8)（B）款中所述的非使用状态积分的车企必须进行测试和/或准备工程评估，以证明该技术在车辆整个使用寿命中的耐用性。

2. 数据和信息要求。要求车企必须上交一份有关（a）(8)（B）款中所述非使用状态积分的申请。申请应包含以下内容：

a. 非使用状态技术及其如何在 FTP 和 HWFET 没有包含的条件下减少 CO_2 排放的详细描述。

b. 将应用该技术的一种或多种车型的名单。

c. 选取的测试车辆的详细描述以及解释选取这些车辆的工程评估。

d. （a）(8)（B）款要求的所有测试和/或模拟数据，以及车企在分析中使用的其他数据。

e. 基于车型的非使用状态效果估值和根据预测的应用该技术的车型的销售量计算的减排效果的估值。

f. 工程评估和/或部件耐用性测试数据或证明非使用状态技术部件耐用性的整车测试数据

3. 非使用状态积分申请的审核。收到车企的申请后，执行官员应当操作以下程序：

a. 审核申请的完整性，如果需要加入其他信息，则需在 30 天内通知车企。

b. 审核申请中提供的数据和信息，确定申请是否支持车企估算的积分水平。

4. 非使用状态申请的决定。在收到完整申请的 60 天内，执行官员须以书面形式通知车企，告知通过或者拒绝申请的决定。如果申请被拒，执行官员应当提供拒绝申请的原因。

（D）非使用状态总积分的计算。PC、LDT 和 MDPV 的以 CO_2 g/mi 为单位的非使用状态积分（就近取整到小数点后一位，单位为 g/mi）应当按照以下公式分别计算：

$$总积分（g/mi）＝积分×产量$$

其中：

积分＝按照（a）(8)(A）或（a）(8)(B）款计算的积分值，单位为 g/mi。

产量＝适用的生产并在加州交付销售的、能够获得（a）(8)(A）或（a）(8)(B）款中所述的积分值的非使用状态技术的 PC、LDT 和 MDPV 总数。

(9) 某些全尺寸皮卡积分。依照本（a）(9）款规定，全尺寸皮卡可以得到基于混合动力技术应用或排放性能的额外积分。积分可以按照（a）(9)(A）或（a）(9)(B）款计算，但是不能同时按照两个条款计算。

（A）油电混合动力技术应用积分。按照本（a）(9)(A）款规定，应用油电混合动力技术的全尺寸皮卡可以得到额外积分。使用本（a）(9)(A）款积分的卡车不可再使用（a）(9)(B）款的积分。

1. 轻度油电混合动力以及 2017 – 2021 年款的全尺寸皮卡能够获得 10 g/mi 的积分。要获得此积分，车企必须生产一定量的轻度混合动力全尺寸皮卡，且每个年款中，该类全尺寸皮卡的产量占该车企的全尺寸皮卡总产量的百分比不小于下表中规定的百分比。

年款	全尺寸皮卡最小百分比要求
2017	30%
2018	40%
2019	55%
2020	70%
2021	80%

2. 强油电混合动力以及 2017 – 2025 年款的全尺寸皮卡能够获得 20 g/mi 的

积分。要获得此积分，车企必须生产一定量的强混合动力全尺寸皮卡，且每个年款中，该类卡车的产量占该车企的全尺寸皮卡总产量的百分比不小于10%。

（B）减排性能积分。 2017－2021年款的全尺寸皮卡，如果碳类尾气排放小于根据（a）(1)(B)款计算的适用目标值，就能够获得额外积分。使用本（a）(9)(B)款中的积分的卡车不能使用（a）(9)(A)款中的积分。

1. 一个年款中，碳类尾气排放小于或等于（a）(1)(B)款计算的适用目标值乘以0.85（就近取整，单位为g/mi），大于（a）(1)(B)款计算的适用目标值乘以0.80（就近取整，单位为g/mi）的全尺寸皮卡可以得到10 g/mi的积分。如果有资格获得此积分的卡车的碳类尾气排放不超过第一次获得此积分的年款的排放量，则该卡车以后的年款直到2021年款可以继续获得此积分。要获得此积分，车企必须生产一定量的符合（a）(9)(B)1款排放要求的全尺寸皮卡，且每个年款中，该类全尺寸皮卡的产量占该车企的全尺寸皮卡总产量的百分比不小于下表中规定的百分比。

年款	全尺寸皮卡最小百分比要求
2017	15%
2018	20%
2019	28%
2020	35%
2021	40%

2. 一个年款中，碳类尾气排放小于或等于（a）(1)(B)款计算的适用目标值乘以0.80（就近取整，单位为g/mi）的全尺寸皮卡可以得到20 g/mi的积分。如果获得此积分的卡车的碳类尾气排放不超过第一次获得此积分的年款的排放量，则该卡车可以在以后最多5年继续获得此积分。此积分对2025年后的所有年款失效。要获得此积分，车企必须生产一定量的符合（a）(9)(B)1款规定的排放要求的全尺寸皮卡，且每个年款中，该类全尺寸皮卡的产量占该车企的全尺寸皮卡总产量的百分比不小于10%。

（C）全尺寸皮卡总积分的计算。 按照CO_2 g/mi计算的有资格获得积分的全尺寸皮卡总积分（就近取整，单位为g/mi）应当按照以下公式计算：

$$总积分（克／英里）= (10 \times 产量_{10}) + (20 \times 产量_{20})$$

其中

产量$_{10}$ = 按照（a）(9)(A)和（a）(9)(B)款计算的积分值为10 g/mi的、生产

并在加州交付销售的全尺寸皮卡总数

产量$_{20}$ = 按照（a）(9)(A) 和（a）(9)(B) 款计算的积分值为 20 g/mi 的、生产并在加州交付销售的全尺寸皮卡总数

（10）温室气体在用合规标准。在用尾气 CO_2 排放标准应当是根据（a）(5)(A) 款计算的某一年款足迹面积的城市/公路组合排放值乘以 1.1，并就近取整，单位为 g/mi。对于多燃料车辆，应当单独计算该车辆可以使用的每一种燃料的值。这些标准适用于《加州乘用车、轻型卡车和中型车辆 2015 及后续年款尾气排放标准和测试程序以及 2017 及后续年款温室气体排放标准和测试程序》中有关车企实施的在用检测。

（11）2022 – 2025 年款标准中期审查。执行官员应进行中期审查，重新评估车辆技术状态，以确定是否需要对 2022 – 2025 年款标准的强度作出调整。加州的中期审查将与 40 CFR §86.1818 – 12（h）款中规定的 EPA 中期评审协调进行。

（b）温室气体积分/借记计算。2012 – 2016 年款全国温室气体规划中得到的积分不适用于加州的温室气体规划。2012 – 2016 年款全国温室气体规划中得到的借记不适用于加州的温室气体规划。

（1）PC、LDT 和 MDPV 温室气体积分计算。

（A）车辆平均 CO_2 值小于对应年款平均 CO_2 目标的车企，其对应年款可获得积分，单位为 g/mi。车辆平均 CO_2 值大于对应年款平均 CO_2 目标的车企，其对应年款可获得借记，单位为 g/mi。车企应当按照以下公式分别计算 PC 以及 LDT 和 MDPV 的联合温室气体积分和温室气体借记：

CO_2 积分/借记总额 = （CO_2 标准 – 车企平均 CO_2 值）×生产并在加州交付销售的车辆总数，包括 ZEV 和 HEV）。

其中：CO_2 标准 = 根据（a）(1)(C) 条款中确定的年款适用标准；

车企平均 CO_2 值 = 根据第（a）(5) 条款计算出的平均值

（B）车企某年款产生的温室气体积分/借记应是 CO_2 积分/借记的总值和适用下述积分/借记的总值。车企须分别计算、保持并报告 PC 产品线、LDT 和 MDPV 温室气体积分。

1. 根据（a）(6) 条款获得的车辆空调泄漏积分；

2. 根据（a）(7) 条款获得的车辆空调效率积分；

3. 根据（a）(8) 条款获得的车辆非使用状态技术积分。

4. 根据（a）(2)(D) 条款下获得的 CO_2 当量借记总额。

（2）若 2017 及后续年款的平均温室气体值高于适用当年年款的平均 CO_2 标准，则车企将获得与上述等式中确定的负积分等值的以 g/mi 为计量单位的温室气体借记

总额。对于 2017 及后续年款中，PC 以及 LDT 和 MDPV 所获得的温室气排放值应合并计总。得出的数值为车企该年款总计获得温室气体积分或者借记总额。

（3）温室气体借记抵消程序。

（A）车企可通过获得温室气体排放积分，或者向执行官提交前期获得的或者从其他车企购买的等值的温室气体积分，来抵消等额的温室气体借记总额。车企必须在获取借记后的 5 个年款内抵消其 PC、LDT 和 MDPV 积累的温室气体借记总额。若车企未能在规定的时间内抵消所有的排放借记总额，则根据《健康和安全法》第 43211 条的规定，车企将因销售不符合州理事会采纳的排放标准的新车型而受到民事惩罚。在规定的时间内未能抵消所有的排放借记总额，则诉讼理由也将形成。对于选择第（a）(5)（D）条款下路径 2 的车企，其排放借记总额在加州、哥伦比亚特区和受第（a）(1）条款下车辆平均温室气体目标约束的其他各州内应分别计算。除生产并在加州交付销售的车辆总数（包括 ZEV 和 HEV），在哥伦比亚特区和其他各州分别计算外，各州的排放借记总额将根据第（b）(1）和（b）(2）条款中的公式计算。

根据《健康和安全法》第 43211 条的规定，不符合州理事会排放标准的 PC 的数量确定方法为：借记产生第一年年款的加州该年款车辆的 g/mi 温室气体借记总额除以 PC 平均温室气体目标。根据《健康和安全法》第 43211 条的规定，不符合州理事会排放标准的 LDT 和 MDPV 的数量确定方法为：借记产生第一年年款的车辆的 g/mi 温室气体借记总额除以 LDT 和 MDPV 平均温室气体目标。

（B）2017 及后续年款获得的温室气体排放积分在获得积分开始至第 5 个年款内保留全值，但从第 6 个年款开始之后，若不使用，将失去价值。

（4）使用温室气体排放积分抵消车企的 ZEV 目标。

（A）在某一确定的年款内，车企在抵消所有的温室气体借记总额之后剩余的温室气体积分，根据引入 1962.2 中作为参考的《加州乘用车、轻型卡车和中型车辆 2018 及后续年款零排放车和混合动力电动车尾气排放标准和测试程序》中的条款，可用来抵消该年款的 ZEV 目标。

（B）车企使用的任何温室气体积分在用来抵消 ZEV 目标后，将不再具有履行 1961.3 章节的价值。

（5）2012-2016 年款国家温室气体规划下获得的积分或借记总额，对于履行 1961.3 章节将失去价值。

（c）2017-2025 年款国家温室气体规划可选达标方式。

2017-2025 年款期间，车企可通过证明符合 2017-2025 年款国家温室气体规划的规定，来达到本 1961.3 的规定，具体如下：

（1） 选择符合本 1963.1 规定的车企，在年款规定生效开始之前，或者按照 1961.3（a）和（b）的规定，必须以书面的形式告知执行官员其选择；

（2） 车企必须按照 40 CFR §86.1865-12 及《加州乘用车、轻型卡车和中型车辆 2015 及后续年款尾气排放标准和测试程序》规定的报告要求，向加州空气资源委员会提交其向国家环境保护局提交的所有数据，以证明其符合 2017-2025 年款国家温室气体规划。所有数据必须在车企收到国家环境保护局确定合规通知后的 30 天内提交；

（3） 车企必须向执行官员提供其在加州、哥伦比亚特区和其他根据联邦《清洁空气法》（42 U.S.C. §7507）第 177 章采纳加州温室气体排放标准的各州生产和交付销售的每一年款/足迹面积组合的车辆数量、平均 CO_2 标准、已计算出的每一年款 CO_2 平均值以及计算平均 CO_2 值所使用的全部数值。

（d） 测试程序本条中确定是否符合排放标准的认证要求和测试程序可参见 1961.2 引用的《乘用车、轻型卡车和中型车辆 2015 及后续年款加州尾气排放标准和测试程序以及 2017 及后续年款温室气体排放标准和测试程序》。对于混合动力车，本条中确定是否符合排放标准的认证要求和测试程序可参见 1961.2 引用的《加州乘用车、轻型卡车和中型车辆 2009-2017 年款零排放车、混合动力电动车尾气排放标准和测试程序》，或 1962.2 节中的《加州乘用车、轻型卡车和中型车辆 2018 及后续年款零排放车和混合动力电动车尾气排放标准和测试程序》。

（e） *缩写词*。在 1961.3 中使用下述缩写词：

"CFR" 指联邦法规（美国）。

"CH_4" 指甲烷。

"CO_2" 指二氧化碳。

"FTP" 指联邦测试程序。

"GHG" 指温室气体。

"g/mi" 指克/英里。

"GVW" 指车辆总重量。

"GVWR" 指车辆额定总重量。

"GWP" 指全球变暖潜能值。

"HEV" 指混合动力电动车。

"HWFET" 指公路燃油经济性测试（HWFET；40 CFR 600 B 章）。

"LDT" 指轻型卡车。

"LVW" 指满载重量。

"MDPV" 指中型乘用车。

"mg/mi" 指毫克/英里。

"MY" 指年款。

"NHTSA" 指国家公路交通安全管理局。

"N_2O" 指氧化亚氮。

"ZEV" 指零排放车。

（f）本节中的相关定义。下述定义适用于本节1961.3：

（1）"空调直接排放"指机动车辆空调系统制冷剂排放。

（2）"主动空气动力改进技术"指在达到一定速度时可激活的技术，可提高空气动力效率至少3%以上，同时维持车辆的其他属性或者功能。

（3）"主动车厢通风设备"指将车厢内的热空气智能排放到车辆外的设备。

（4）"变速器主动预热"指利用排气系统内置的热交换器收集余热以加热传动液至可运作的温度，降低因传动液造成的热量损失，如摩擦和流体粘性造成的损失，以提高整体传动效率。

（5）"发动机主动预热"指利用排气系统的余热加热发动机目标零件以减少发动机摩擦热损和保证燃油闭环回路控制系统更快地启动设备，可加快从低温运转至正常运转的过渡，降低 CO_2 的排放。

（6）"主动座椅通风设备"指吸收人体与座椅接触表面的热量并将其排放到远离座椅位置的设备。

（7）"减少废能损耗的鼓风控制"指通过控制风扇和鼓风速度，避免启动电阻原件，降低发动机电压的方法。

（8）"默认空气内循环模式"指当驾驶员或者自动空气控制系统启动空调系统时，控制空调系统空气来源的默认状态，可以从外空气循环模式转换至内循环模式（例如，蒸发器正在除热），为提高能见度需要除湿情况下（即除雾模式）除外。配备有内部空气质量感应器（即湿度感应器、CO_2 感应器）的车辆可以自动调节空气来源以保证车厢内空气的新鲜和防止持续内循环模式下汽车窗玻璃形成雾气。在车辆运行的任何时候，驾驶员可手动选择非空气循环模式但是系统在车辆后来运行中（即再次启动时）默认为空气内循环模式。空气控制系统可延迟转换为内循环模式直到车厢内温度低于外部空气温度，此时须启动空气内循环模式。

（9）"电加热循环泵"指安置在配备有起停系统的车辆或混合动力电动车或插电式混合动力电动车内的泵系统，在起停阶段发动机熄火后，仍可保证热的冷却液循环流过发热器芯。该系统必须校准，以保证周围温度达到30华氏摄氏度时，发动机熄火时间可长至1分钟或者更长时间。

（10）"应急车辆"指在美国专门作为救护车或救护车 – 灵车两用车的机动车

辆，以及美国政府或者州政府用来执法的车辆。

（11）"发动机热回收系统"指可以捕获通过排气系统或者散热器损耗的热量并将热量转换为符合车辆电力要求的电能的系统。该系统容量至少为100W，才能获得0.7 g/mi 的积分。容量每增加 100W 则增加 0.7 g/mi 的积分。

（12）"发动机起停技术"指车辆休息时自动关闭发动机，驾驶员踩下油门或者松开刹车时又自动启动发动机的技术。

（13）"环境保护局车辆模拟工具"指 40 CFR §86.1 和《关于环境保护局2017及后续年款国家温室气体规划规则制定建议的通知》中引用的及 76 联邦法规74854，75357（2011 年 12 月 1 日）提出的"环境保护局车辆模拟工具"。

（14）"执行官员"指加州空气资源委员会的执行官员。

（15）"足迹面积"指轮距（就近舍入到 1/10 英寸）和轴距（英寸测量，就近舍入到 1/10 英寸）的平均值，然后被 144 除，得出的结果舍入到 1/10 平方英尺，其中平均轮距是前后轮距的平均值，前后轮距用英寸测量，舍入到 1/10 英寸。

（16）"联邦测试程序"或"FTP"指 40 CFR，第 86 章，B 节（40 CFR, Part 86, Subpart B），经《加州乘用车、轻型卡车和中型车辆 2015 及后续年款尾气排放标准和测试程序以及 2017 及后续年款温室气体排放标准和测试程序》修订。

（17）"全尺寸皮卡"是一种轻型卡车，既有一个封闭的乘客车厢（驾驶室）又有一个敞篷的货物车厢，规格如下：

1. 货物车厢在两轮之间最小宽度为 48 英寸，通过测量驾驶室可通行左右两边边界线的横向最小距离得出，不包括过渡弧线距离、凸起部分和凹陷或者容器（如果存在）部分。敞篷货厢指车辆没有一个固定的篷或者遮盖层。根据本标准，可拆卸式车篷也视为"敞篷"。

2. 敞篷货厢长度至少为 60 英寸，货厢长度指从车身顶部测量出的车厢长度和从车厢底板位置测量出的货厢长度中的较小值，前者是指在车身顶部位置高度沿着车辆中线从车厢内部最前面至车厢后门的内侧的纵向长度，后者是指在车厢底板位置沿着车辆中线从车厢内部最前面至车厢后门的内侧的纵向长度。

3. 牵引能力最低为 5 000 磅，牵引能力指从综合额定总重中减去额定车辆总重；或者有效载荷能力至少为 1 700 磅，有效载荷能力指车辆额定总重减去车辆净重。

（18）"温室气体"指以下气体：二氧化碳、甲烷、氧化亚氮和氢氟碳化物。

（19）"全球变暖潜能值"指在 100 年的时间跨度内制冷剂排放导致的全球变暖潜能值，在 2007 年政府间气候变化委员会（IPCC）有解释：《气候变化 2007——自然科学基础》S. Solomon 等编辑，政府间气候变化委员会对于气候变化的第四次评估报告第一工作小组的成果，剑桥大学出版社，剑桥，英国和纽约，纽约，美国，ISBN 0 - 521 - 70596 - 7 如果政府间气候变化委员会第四次评估报告无法获得该信息，则依据环境保护局的规定。

（20）"高效外部照明技术"相对于传统的照明系统，车辆安装此外部照明技术系统后可至少减少 60 瓦的电耗。可获得积分的高效照明须安装在下述位置：双闪（parking/position）灯，前后转向信号灯，前后轮廓灯，停车/制动灯（包括中部位置），尾灯，倒车灯，牌照灯。

（21）"改进的冷凝器和/或蒸发器"指安装改进的蒸发器和冷凝器的空调系统的制冷性能相对于标准或者之前年款的设计可提高 10%，该值采用 SAE J2765 规定的基准测试程序，即《机动车空调系统基准测试程序》，得出。SAE J2765 在此处引用。车企必须提交工程分析证明车辆的制冷系统性能相对于标准设计有改进，对照的标准设计零部件是指车企最近生产的适用于同一车辆或者类似车辆的零部件。须对比标准零部件的尺寸参数（例如：排气管结构、厚度、管距、密度）和改进的零部件的尺寸参数，以证明性能的改进。

（22）"轻度混合动力电动车"指具有起停系统和再生制动系统的车辆，根据联邦测试程序，再生制动能量占总制动能量的 15%—75%，再生制动能量的百分比依照 40 CFR §600.108（g）测量和计算。

（23）"车型"指产品品线、发动机和传动装置的特定组合。

（24）"2012 - 2016 年款国家温室气体规划"指适用于 2012 - 2016 年款的新产乘用车、轻型卡车、中型乘用车的国家规划，于 2010 年 4 月 1 日由美国国家环保护局通过（Fed. Reg. 25324，25677（2010 年 5 月 7 日））。

（25）"2017 - 2025 年款国家温室气体规划"指适用于 2017 - 2025 年款的乘用车、轻型卡车、中型乘用车的国家规划，已获得美国国家环境保护局通过，编入 40 CFR Part 86，Subpart S。

（26）"油分离器"指用来分离压缩机中油/制冷剂混合物中夹杂的油，分离率高于 50%，将分离出的油送回压缩室或者送至压缩机进口，或者其他不需要油/制冷剂混合物起润滑作用的压缩设置。

（27）"被动车厢通风设备"指利用对流气流原理将车厢内的热空气排放到车辆外部的设备。

（28）"插电式混合动力电动车"指可外接充电的混合动力车，见《加州乘用车、轻型卡车和中型车辆 2018 及后续年款零排放车和混合动力电动车尾气排放标准和测试程序》中对此的定义。

（29）"减少再加热、外部控制的、固定排量或者气动可变排量压缩机"指传感器的输入信号（例如内部温度控制的状态或设定值，内部温度，蒸发器出口处空气温度，或者制冷剂的温度）转换为电子信号，通过电子信号控制压缩机的开关，以此控制压缩机的输出；该系统可控制蒸发器出口处的空气温度在 41 华氏摄氏度或者更高温度。

（30）"减少再加热、外部控制的、可变排量压缩机"压缩机的排量通过传感器的输入信号（例如内部温度控制的状态或设定值，内部温度，蒸发器出口处空气温度，或者制冷剂的温度）产生的电子信号来实现控制的压缩机系统；该系统可控制蒸发器出口处的空气温度在 41 华氏摄氏度或者更高温度。

（31）"SC03"指 SC03 测试循环，见《加州乘用车、轻型卡车和中型车辆 2015 及后续年款尾气排放标准和测试程序以及 2017 及后续年款温室气体排放标准和测试程序》中的说明。

（32）"太阳能反射涂料"指车漆或者车辆表面涂层的红外线太阳能反射率至少高于 65%，计算方式可参见 ASTM 标准 E903 – 96（即使用积分计算的太阳能吸收、反射、透射率标准测试法，DOI：10. 1520/E0903 – 96（2005 年撤销））或 E1918 – 06（实地水平表面或低坡度表面太阳能反射率标准测试法，DOI：10. 1520/E1918 – 06）或 C1549 – 09（太阳能反射计环境温度下太阳能反射率标准测试法，DOI：10. 1520/C1549 – 09）。这些 ASTM 标准在此处引用参考。

（33）"太阳能顶板"指安装在电动车或插电式混合动力电动车上的太阳能板，可吸收太阳能为车辆电池充电或直接向车辆电动发动机提供相当于至少 50 瓦的额定发电量的能量，从而启动车辆电驱动系统。

（34）"强度油电混合动力车"指具有起停系统和再生制动系统的车辆，根据联邦测试程序，再生制动能量占总制动能量 75% 以上，再生制动能量依照 40 CFR §600. 108（g）测量和计算。

（35）"子配置"指在同等测试重量、道路负荷马力和其他美国国家环境保护局所认证的运行特征或参数下车辆配置内部的特殊组合。

（36）"US06"指指 SC06 测试循环，见《加州乘用车、轻型卡车和中型车辆 2015 及后续年款尾气排放标准和测试程序以及 2017 及后续年款温室气体排放标准和测试程序》中的说明。

（37）"最差配置"指每一个测试组中根据（a）(5)计算出的最高 CO_2 当量值的车辆配置。

（g）可分割性。本节中的每条规定都是可分割的，如本节的某一条规定被认定为无效，其余的所有规定仍然继续完全有效。

注释：文献引用：《健康和安全法》第 39500，39600，39601，43013，43018，

43018.5，43101，43104 和 43105 节。文献参考：《健康和安全法》第 39002，39003，39667，43000，43009.5，43013，43018，43018.5，43100，43101，43101.5，43102，43104，43105，43106，43204，43205 和 43211 节。

§ 1976　机动车辆燃料挥发排放标准和测试程序

（**a**）【1970 – 1977 年款乘用车（PC）和轻型卡车（LDT）燃料挥发排放标准；不做阐述】

（**b**）（1）遵守本条款尾气排放标准的 1978 及后续年款汽油燃料车辆，1983 及后续年款液化石油气燃料车辆，以及 1993 及后续年款乙醇燃料机动车辆和混合动力车辆，不包括石油燃料柴油车辆、压缩天然气燃料车辆和混合动力车辆（拥有封闭的燃料系统，没有挥发排放）以及摩托车，它们的挥发排放不能超过以下标准。

（**G**）2015 及后续年款机动车辆适用以下挥发排放标准。

1. 车企必须证明所有车辆符合本条款选项 1 或者选项 2 规定的排放标准。

a. 选项 1。根据《加州 2001 及后续年款机动车辆蒸发排放标准和测试程序》（1976（c）章节引用）进行测试，2015 及后续年款机动车辆的蒸发排放不应超过：

车辆类型	碳氢化合物[1] 排放标准[2]		
	运转损失 (g/mi①)	三个白昼 + 热浸和两个白昼 + 热浸	
		整车（克/测试）	燃料[3] （克/测试）
PC	0.05	0.35	0
LDT 额定总重量（GVWR）≤ 6 000 磅（lbs）	0.05	0.5	0
LDT GVWR 在 6001 – 8 500 lbs	0.05	0.75	0
中型乘用车（MDPV）	0.05	0.75	0
中型车辆（MDV） GVWR 在 8501 – 14000 lbs 之间	0.05	0.75	0
重型车辆（HDV） GVWR 超过 14000 lbs	0.05	0.75	0

[1] 乙醇燃料车辆的有机碳氢化合物当量。

[2] 对于符合以上标准的所有车辆，使用寿命应是 15 年或者 150 000 英里，以先到者为准。对于未根据 1961 章节，第 13 主题，加州法规使用底盘测功机进行尾气测试的车辆，应根据申请者提

---

① g/mi，克/英里

交的车辆系统和数据进行工程评估。

[3]车企可在征得执行官员事先同意的情况下选择另一种可替代的测试计划证明合规性，而无需证明遵守燃料排放标准（通过三个白昼 + 热浸以及两个白昼 + 热浸测试，0.0 克/测试）。

b. 选项 2。根据《加州 2001 及后续年款机动车辆挥发排放标准和测试程序》（1976（c）章节引用）进行测试，2015 及后续年款机动车辆的挥发排放不应超过：

车辆类型	碳氢化合物[1]排放标准[2]		
	运转损失 （g/mi）	最高整车 + 热浸[3,4,5] （克/测试）	过滤罐泄漏[6] （克/测试）
PC；LDT GVWR ≤ 6 000 lbs，满载重量（LVW）在 0 – 3750 lbs 之间	0.05	0.3	0.02
LDT GVWR ≤ 6 000 lbs，LVW 在 3 751 – 5 750 lbs 之间	0.05	0.4	0.02
LDT GVWR 在 6001 – 8 500 lbs 之间；中型 PC	0.05	0.5	0.02
MDV GVWR 在 8501 – 14000 lbs 之间；HDV GVWR 超过 14000 lbs	0.05	0.6	0.03

[1]乙醇燃料车辆的有机的碳氢化合物当量。

[2]对于符合运转损失和最高整车白昼 + 热浸排放标准的车辆，使用寿命应为 15 年或者 150 000 英里，以先到者为准。对于未根据 1961 章节，13 主题，加州法规使用底盘测功机进行尾气排放测量的车辆，应根据申请者提交的车辆系统和数据进行工程评估。过滤罐泄漏排放标准没有使用寿命方面的要求。

[3]车企可以选择整车白昼 + 热浸排放测试的最高值（三天白昼 + 热浸测试和两天白昼 + 热浸测试）来判断是否合规。

[4]每个排放标准种类内最高整车白昼 + 热浸排放标准的平均选项。车企可选择使用平均碳氢化合物排放值以达到最高整车白昼 + 热浸排放标准。为此，车企针对所有排放标准种类必须使用平均选项，并且针对每一排放标准种类计算单独的平均碳氢化合物排放值。排放标准种类如下：（1）PC 和 LDT（0 < GVWR ≤ 6 000 lbs，0 < LVW < 3750 lbs）；（2）LDT（0 < GVWR ≤ 6 000 lbs，3 751 lbs < LVW < 5 750 lbs）；（3）LDT（6001 lbs < GVWR < 8 500 lbs）和 MDPV；（4）MDV 和 HDV。每个排放标准种类的平均碳氢化合物排放值根据以下公式计算：

$$\sum_{i=1}^{n}\left[\text{所属挥发类别的车辆数量}\,i \times \text{挥发类别排放限值}\,i\right] \div \sum_{i=1}^{n} \text{所属挥发类别的车辆数量}\,i$$

其中，

"n" = 某一年款排放标准种类下，车企符合选项 2 的挥发类别。

"所属挥发类别的车辆数量" = 生产并在加州交付销售的，蒸发类别的车辆。

数量。

"挥发类别排放限值"=车企针对挥发类别选择的数值,该数值通过所有测试所得,作为该排放类别的排放标准,而不是本条款 1976(b)(1)(G)1.b.的排放标准。PC 所属挥发类别排放限值不应超过 0.5 克/测试,LDT(GVWR≤6 000)不应超过 0.650 克/测试;LDT(GVWR 在 6001 – 8 500 之间)不应超过 0.9 克/测试;MDPV、MDV 和 HDV 不应超过 1.0 克/测试。除此之外,挥发类别排放限值增量为 0.025 克/测试。

⁵ 计算平均选项中碳氢化合物积分和借记。

(1)计算碳氢化合物积分和借记。针对某一年款排放标准种类,车企须根据以下公式计算碳氢化合物积分和借记:

【(某排放标准种类下适用碳氢化合物排放标准)—(车企在该排放标准种类下的平均碳氢化合物排放值)】×涉及的车辆数量

"涉及的车辆数量"=某一年款排放标准种类下,平均选项涉及的排放类别内的所有生产并在加州交付销售的车辆数量。

若得出的数值为负,则构成碳氢化合物借记,正值代表积分,车企可在当年款累计。某一年款获得的碳氢化合物积分将保留全部价值直至第 5 个年款。在第 6 个年款开始时,积分失效。

(2)抵消碳氢化合物借记的程序。车企可以利用借记产生之后的三个年款内的对应排放标准种类的碳氢化合物积分来抵消借记值。如果总借记值在三个年款内未被抵消,车企将会依据《健康和安全法》第 43211 节中的民事处罚条例受到制裁,该条例适用于那些所售新机动车未达到空气资源委员会规定的适用排放标准的车企。当所欠积分截至规定时间依然未能偿清,则有正当理由对该车企进行相应的处罚。根据《健康和安全法》第 43211 节,未满足州排放标准的车辆数量等于,某年款内排放种类下碳氢化合物借记总值除以借记产生的年款内适用的碳氢化合物排放标准。

除此之外,在三个年款补足借记期限结束时,为了补足所欠积分:(1)碳氢化合物积分可以在 PC、LDT(GVRW≤6 000 lbs, 0 < LVW < 3750 lbs)、LDT(GVWR≤6 000 lbs, 3 751 lbs < LVW < 5 750 lbs)之间交换;(2)碳氢化合物积分可在 LDT(6001 lbs < GVWR < 8 500 lbs)、MDPV、MDV 和 HDV 之间交换。

⁶机动车挥发过滤罐泄露排放。是否符合挥发过滤罐泄露排放标准需要根据过滤罐排放测试程序来判断,可参见《加州 2001 及后续年款机动车辆挥发排放标准和测试程序》(1976(c)中引用),并且在稳定的过滤罐系统中展示。如果车辆拥有不完整的补给燃料过滤罐系统,则可免于遵守过滤罐泄露排放标准。

2. 渐进式表。 根据以下表格所述,针对每一年款,车企需证明其至少一定比例的车辆符合 1976(b)(1)(G)1.a. 或者 1976(b)(1)(G)1.b. 的挥发排放标准。根据本条款 1976(b)(1)(G)2,车企所产机动车包含生产并在加州交付销售的,并且遵守 1976(b)(1)(G)1 规定的排放标准的车辆。根据以下渐进式安排不符合这些标准的 2015 – 2022 年款机动车，应遵守 2004 – 2014 年款机动车标准，详见 1976 （b）（1）（F）。

年　　款	最小比例[1,2]
2015，2016，2017	2012，2013，2014 年款符合 1976 （b）（1）（E） 条款的车辆平均值[3,4]
2018，2019	60
2020，2021	80
2022 及后续年款	100

[1] 仅针对 2018 – 2022 年款，车企可选择使用替代的渐进式表来达标。替代的渐进式表必须在最后一个年款结束时（2022 年款）达到相应的合规量。合规量等于每一年款满足新标准的车辆比例（基于车企预计所有车辆的销售量），乘以渐进式表最后一个年款（包含）之前执行的年份数目，然后将各年结果加总得出一个总数。上述表格中 5 年的总和根据以下公式计算：

（60 × 5 年）+（60 × 4 年）+（80 × 3 年）+（80 × 2 年）+（100 × 1 年）= 1 040。相应的，替代渐进式表规定的总数也是 1040。执行官员需要考虑任何可接受的渐进式表，其在最后一个年款结束时的合规量等于或者大于上文所述的总数。

[2] 小型车企不必遵守上述渐进式表。相反，它们需证明 2022 及后续年款 100% 的车辆满足 1976 （b）（1）（G） 1.a 或者 1976 （b）（1）（G） 1.b 规定的挥发排放标准。

[3] 2015、2016 和 2017 年款车辆比例的平均值将用来决定是否合规。

[4] 2015、2016 和 2017 年款规定的最小比例等于，车企 2012、2013 和 2014 年款符合 1976 （b）（1）（E） 排放标准的车辆比例的平均值。为了计算这个平均值，车企须使用生产并在加州交付销售的 2012、2013 和 2014 年款车辆的比例。车企在计算平均值时可使用预测的销售量而非实际的销售量。

3. 符合零燃料挥发排放标准的 2014 年款所属挥发类别的结转。车企可以结转符合 1976 （b）（1）（E） 规定的零燃料挥发排放标准的 2014 年款机动车至 2018 年款，并且认为符合 1976 （b）（1）（G） 1 规定的目标。对所有符合结转条款的机动车辆，在决定车辆使用寿命期间在用合规标准时应使用 1976 （b）（1）（E） 规定的排放标准。如果车企选择遵守平均选项最高整车白昼 + 热浸排放标准，以下类别排放限值对应这些挥发类别，以计算车企平均碳氢化合物排放值。

车辆类型	最高整车白昼 + 热浸 （克/测试）
PC	0.300
LDT （GVWR ≤ 6 000 lbs；0 < LVW < 3 750 lbs）	0.300
LDT （GVWR ≤ 6 000 lbs；3 751 < LVW < 5 750 lbs）	0.400
LDT （6 001 < GVWR < 8 500 lbs）	0.500

4. 合并条款。以下合并条款适用于 1976 （b）（1）（G） 1.b 最高整车白昼 +

热浸排放标准的平均选项以及 1976（b）（1）（G）2. 规定的渐进式目标。

a. 根据 1976（b）（1）（G）1.b 规定的平均选项，车企针对每一年款必须基于以下两个选项之一证明合规。

合并选项 1：符合 1976（b）（1）（G）1.b 规定的加州挥发排放标准的，生产并在加州交付销售的，PC、LDT、MDPV、MDV 和 HDV 的总和；

合并选项 2：符合 1976（b）（1）（G）1.b 规定的加州挥发排放标准的，生产并在加州、哥伦比亚特区以及依据联邦《清洁空气法案》（42 U.S.C. § 7507）第 177 章遵守 1976（b）（1）（G）1 规定的加州挥发排放标准的所有各州，交付销售的 PC、LDT、MDPV、MDV 和 HDV 的总和。

b. 根据 1976（b）（1）（G）2. 规定的渐进式目标，车企必须基于以下两个选项之一证明合规。

合并选项 1：符合 1976（b）（1）（G）1 规定的加州挥发排放标准的，生产并在加州交付销售的 PC、LDT、MDPV、MDV 和 HDV 的总数。

合并选项 2：符合 1976（b）（1）（G）1 规定的加州挥发排放标准的，生产并在加州、哥伦比亚特区以及依据联邦《清洁空气法案》（42 U.S.C. § 7507）第 177 章遵守 1976（b）（1）（G）1 规定的加州挥发排放标准的所有各州，交付销售的 PC、LDT、MDPV、MDV 和 HDV 的总和。

c. 选择合并选项 2 的车企必须在适用年款开始之前书面告知执行官员这一选择，否则必须遵守合并选项 1。一旦车企选择合并选项 2，就须遵守，除非车企在适用年款开始之前书面告知执行官员选择选项 1。

d. 当车企在某一年款证明符合合并选项 2 时，1976（b）（1）（G）条款中"在加州"意指加州、哥伦比亚特区以及依据联邦《清洁空气法案》（42 U.S.C. § 7507）. 遵守加州挥发排放标准的所有各州。

e. 选择合并选项 2 的车企必须向执行官员提交生产并在哥伦比亚特区以及第 177 章所列各州交付销售的，每个挥发类别对应的车辆数量。

5. 2014 年款机动车的可选认证。车企可选择证明其 2014 年款机动车符合 1976（b）（1）（G）1 规定的挥发排放标准。

注释：文献引用：《健康和安全法》第 39500，39600，39601，39667，43013，43018，43101，43104，43105，43106 和 43107 节。文献参考：39002，39003，39500，39667，43000，43009.5，43013，43018，43100，43101，43101.5，43102，43104，43105，43106，43107，43204 和 43205 节。

第三部分
美国加州零排放车和低排放车法规原文

The California Zero-Emission Vehicle Regulations

§ 1962. 1 Zero-Emission Vehicle Standards for 2009 through 2017 Model Year Passenger Cars, Light-Duty Trucks, and Medium-Duty Vehicles.

(a) *ZEV Emission Standard*. The Executive Officer shall certify new 2009 through 2017 model year passenger cars, light-duty trucks and medium-duty vehicles as ZEVs if the vehicles produce zero exhaust emissions of any criteria pollutant (or precursor pollutant) under any and all possible operational modes and conditions.

(b) *Percentage ZEV Requirements*.

(1) *General Percentage ZEV Requirement*.

(A) *Basic Requirement*. The minimum percentage ZEV requirement for each manufacturer is listed in the table below as the percentage of the PCs and LDT1s, and LDT2s to the extent required by subdivision (b)(1)(C), produced by the manufacturer and delivered for sale in California that must be ZEVs, subject to the conditions in this subdivision 1962. 1(b). The ZEV requirement will be based on the annual NMOG production report for the appropriate model year.

Model Years	Minimum ZEV Requirement
2009 through 2011	11 %
2012 through 2014	12 %
2015 through 2017	14 %

（B）*Calculating the Number of Vehicles to Which the Percentage ZEV Requirement is Applied*. For purposes of calculating a manufacturer's requirement in subdivision 1962. 1(b)(1) for model years 2009 through 2017, a manufacturer may use a three year average method or same model year method, as described below in sections 1. and 2. A manufacturer may switch methods on an annual basis. This production averaging is used to determine ZEV requirements specified in subdivision 1962. 1(b)(1)(A) only, and has no effect on a manufacturer's size determination, specified in section 1900. In applying the ZEV requirement, a PC, LDT1, or LDT2, that is produced by one manufacturer (e. g. , Manufacturer A), but is marketed in California by another manufacturer (e. g. , Manufacturer B) under the other manufacturer's (Manufacturer B) nameplate, shall be treated as having been produced by the marketing manufacturer (Manufacturer B).

1. For the 2009 through 2011 model years, a manufacturer's production volume of PCs and LDT1s, and LDT2s as applicable, produced and delivered for sale in California will be based on the three-year average of the manufacturer's volume of PCs and LDT1s, and LDT2s as applicable, produced and delivered for sale in California in the 2003 through 2005 model years. As an alternative to the three-year averaging of prior year production described above, a manufacturer may elect to base its ZEV obligation on the number of PCs and LDT1s, and LDT2s, as applicable, produced by the manufacturer and delivered for sale in California that same model year.

2. For 2012 through 2017 model years, a manufacturer's production volume for the given model year will be based on the three-year average of the manufacturer's volume of PCs and LDTs, produced and delivered for sale in California in the prior fourth, fifth and sixth model year [for example, 2013 model year ZEV requirements will be based on California production volume of PCs and LDTs, for the 2007 to 2009 model years, and 2014 model year ZEV requirements will be based on California production volume of PCs and LDTs, for the 2008 to 2010 model years]. As an alternative to the three-year averaging of prior year production described above, a manufacturer may elect to base its ZEV obligation on the number of PCs and LDTs, produced by the manufacturer and delivered for sale in California that same model year.

（C）*Phase-in of ZEV Requirements for LDT2s*. Beginning with the ZEV requirements for the 2009 model year, a manufacturer's LDT2 production shall be included in determining the manufacturer's overall ZEV requirement under subdivision (b)(1)(A) in the increasing percentages shown in the table below.

2009	2010	2011	2012 +
51%	68%	85%	100%

(**D**) ***Exclusion of ZEVs in Determining a Manufacturer's Sales Volume.*** In calculating, for purposes of subdivisions 1962. 1(b)(1)(B) and 1962. 1(b)(1)(C), the volume of PCs, LDT1s, and LDT2s that a manufacturer has produced and delivered for sale in California, the manufacturer shall exclude the number of ZEVs produced by the manufacturer, or by a subsidiary in which that manufacturer has a greater than 50 percent ownership interest, and delivered for sale in California.

(**2**) ***Requirements for Large Volume Manufacturers.***

(**A**) ***Primary Requirements for Large Volume Manufacturers through Model Year 2011.***

In the 2009 through 2011 model years, a manufacturer must meet at least 22. 5 percent of its ZEV requirement with ZEVs or ZEV credits generated by such vehicles, and at least another 22. 5 percent with ZEVs, AT PZEVs, or credits generated by such vehicles. The remainder of the manufacturer's ZEV requirement may be met using PZEVs or credits generated by such vehicles.

(**B**) ***Alternative Requirements for Large Volume Manufacturers through Model Year 2011.***

1. ***Minimum Floor for Production of Type III ZEVs.***

a. [***Reserved***].

b. ***Requirement for the 2009 – 2011 Model Years.*** A manufacturer electing the alternative compliance requirements during model years 2009 through 2011 must produce ZEV credits equal to 0. 82 percent of the manufacturer's average annual California sales of PCs and LDT1s, and LDT2s, as applicable, over the three-year period from model years 2003 through 2005, through production, delivery for sale, and placement in service of ZEVs, other than NEVs and Type 0 ZEVs, using the credit substitution ratios for each ZEV Type compared to a Type III prescribed in the table below, or submit an equivalent number of credits generated by such vehicles.

ZEV Types	Credit Substitution Ratio Compared To A Type III ZEV
Type I	2
Type I. 5	1. 6
Type II	1. 33
Type IV	0. 8
Type V	0. 57

i. Manufacturers may use credits generated by 1997 – 2003 model year ZEVs that qualify for an extended service multiplier under subdivision 1962. 1(f) for a year during calendar years 2009 – 2011, provided that 33 years of such a multiplier will equal 4 ZEV credits.

c. [*Reserved*].

d. [*Reserved*].

e. [*Reserved*].

f. *Exclusion of Additional Credits for Transportation Systems.* Any additional credits for transportation systems generated in accordance with subdivision 1962. 1(g)(5) shall not be counted towards compliance with this subdivision 1962. 1(b)(2)(B)1. b.

g. *Carry-over of Excess Credits.* ZEV credits generated from excess production in model years 2005 through 2008 may be carried forward and applied to the 2009 through 2011 minimum floor requirement specified in subdivision 1962. 1(b)(2)(B)1. b. provided that the value of these carryover credits shall be based on the model year in which the credits are used. Beginning with the 2012 model year, these credits may no longer be used to meet the ZEV requirement specified in subdivision 1962. 1(b)(2)(B)1. b. ; they may be used as TZEV, AT PZEV, or PZEV credits. ZEV credits earned in model year 2009 through 2011 would be allowed to be carried forward for two years for application to the ZEV requirement. For example, ZEV credit earned in the 2010 model year would retain full flexibility through the 2012 model year. Starting 2013 model year, that credit could only be used as TZEV, AT PZEV, or PZEV credits, and could not be used to satisfy the ZEV credit obligation, which may only be satisfied with credit generated from ZEVs.

h. *Failure to Meet Requirement for Production of ZEVs.* A manufacturer that, after electing the alternative requirements in subdivision 1962. 1(b)(2)(B) for any model year from 2009 through 2011, fails to meet the requirement in subdivision 1962. 1(b)(2)

(B)1. b. by the end of the 2011 model year, shall be treated as subject to the primary requirements in subdivision 1962. 1(b)(2)(A)for the 2009 through 2011 model years.

i. *Rounding Convention*. The number of ZEVs needed for a manufacturer under subdivision 1962. 1(b)(2)(B)1. b. shall be rounded to the nearest whole number.

2. *Compliance with Percentage ZEV Requirements*. In the 2009 through 2011 model years, a manufacturer electing the alternative compliance requirements in a given model year must meet at least 45 percent of its ZEV requirement for that model year with ZEVs, AT PZEVs, TZEVs, or credits generated from such vehicles. ZEV credits generated for compliance with the alternative requirements during any given model year will be applied to the 45 percent which may be met with ZEVs, AT PZEVs, TZEVs, or credits generated from such vehicles, but not PZEVs. The remainder of the manufacturer's ZEV requirement may be met using PZEVs or credits generated from such vehicles.

3. *Sunset of Alternative Requirements after the 2011 Model Year*. The alternative requirements in subdivision 1962. 1(b)(2)(B) are not available after the 2011 model year.

(C) *Election of the Primary or Alternative Requirements for Large Volume Manufacturers for the 2009 through 2011 Model Years*. A manufacturer shall be subject to the primary ZEV requirements for the 2009 model year unless it notifies the Executive Officer in writing prior to the start of the 2009 model year that it is electing to be subject to the alternative compliance requirements for that model year. Thereafter, a manufacturer shall be subject to the same compliance option as applied in the previous model year unless it notifies the Executive Officer in writing prior to the start of a new model year that it is electing to switch to the other compliance option for that new model year. However, a manufacturer that has previously elected the primary ZEV requirements for one or more of the 2009 through 2011 model years may prior to the end of the 2011 model year elect the alternative compliance requirements for the 2009 through 2011 model years upon a demonstration that it has complied with all of the applicable requirements for that period in subdivision 1962. 1 (b)(2)(B)1. b.

(D) *Requirements for Large Volume Manufacturers in Model Years 2012 through 2017*.

1. *2012 through 2014 Requirements*. On an annual basis, a manufacturer must meet the total ZEV obligation with ZEV credits generated by such vehicles, excluding credits generated by NEVs and Type 0 ZEVs, equal to at least 0. 79% of its annual sales, using either production volume determination method described in subdivision 1962. 1 (b) (1) (B). No more than 50% of the total obligation may be met with credits generated from PZEVs. No more than 75% of the total obligation may be met with credits generated from AT PZEVs. No more than 93. 4% may be met with credits generated from TZEVs, Type 0 ZEVs, and NEVs, as limited in subdivision 1962. 1 (g) (6). The entire obligation may be met solely with credits generated from ZEVs.

2. 2015 through 2017 Requirements. On an annual basis, a manufacturer must meet its ZEV obligation with ZEV credits generated by such vehicles, excluding credits generated by NEVs and Type 0 ZEVs, equal to at least 3% of its annual sales, using either production volume determination method described in subdivision 1962. 1 (b) (1) (B). No more than 42. 8% of the total obligation may be met with credits generated from PZEVs. No more than 57. 1% of the total obligation may be met with credits generated from AT PZEVs. No more than 78. 5% may be met with credits generated from TZEVs, Type 0 ZEVs, and NEVs, as limited in subdivision 1962. 1 (g) (6). The entire obligation may be met solely with credits generated from ZEVs.

3. The following table enumerates a manufacturer's annual percentage obligation for the 2012 though 2017 model years if the manufacturer produces the minimum number of credits required to meet its ZEV obligation and the maximum percentage for the TZEV, AT PZEV, and PZEV categories.

Model Years	Total ZEV Percent Requirement	Minimum ZEV floor	TZEVs, Type 0s, or NEVs	AT PZEVs	PZEVs
2012 – 2014	12	0. 79	2. 21	3. 0	6. 0
2015 – 2017	14	3. 0	3. 0	2. 0	6. 0

4. Use of Additional Credits for Transportation Systems. Any additional credits for transportation systems generated from ZEVs in accordance with subdivision 1962. 1 (g) (5) may be used to meet up to one tenth of the portion of the ZEV obligation which must be met with ZEVs, specified in subdivision 1962. 1 (b) (2) (D).

(E) ［Reserved］.

(3) Requirements for Intermediate Volume Manufacturers. For 2009 through 2017 model years, an intermediate volume manufacturer may meet its ZEV requirement with up to 100 percent PZEVs or credits generated by such vehicles. For 2015 through 2017 model years, the overall credit percentage requirement for an intermediate volume manufacturer will be 12%.

(4) *Requirements for Small Volume Manufacturers and Independent Low Volume Manufacturers*. A small volume manufacturer or an independent low volume manufacturer is not required to meet the percentage ZEV requirements. However, a small volume manufacturer or an independent low volume manufacturer may earn and market credits for the ZEVs, TZEVs, AT PZEVs or PZEVs it produces and delivers for sale in California.

(5) [*Reserved*].

(6) [*Reserved*].

(7) *Changes in Small Volume, Independent Low Volume, and Intermediate Volume Manufacturer Status*.

(A) *Increases in California Production Volume*. In 2009 through 2017 model years, if a small volume manufacturer's average California production volume exceeds 4 500 units of new PCs, LDTs, and MDVs based on the average number of vehicles produced and delivered for sale for the three previous consecutive model years, or if an independent low volume manufacturer's average California production volume exceeds 10 000 units of new PCs, LDTs, and MDVs based on the average number of vehicles produced and delivered for sale for the three previous consecutive model years, the manufacturer shall no longer be treated as a small volume, or independent low volume manufacturer, as applicable, and shall comply with the ZEV requirements for intermediate volume manufacturers, as applicable, beginning with the sixth model year after the last of the three consecutive model years.

If an intermediate volume manufacturer's average California production volume exceeds 60 000 units of new PCs, LDTs, and MDVs based on the average number of vehicles produced and delivered for sale for the three previous consecutive model years (i. e., total production volume exceeds 180 000 vehicles in a three-year period), the manufacturer shall no longer be treated as an intermediate volume manufacturer and shall, beginning with the sixth model year after the last of the three consecutive model-years, or in model year 2018 (whichever occurs first), comply with all ZEV requirements for LVMs.

Requirements will begin in the sixth model year, or in model year 2018 (whichever occurs first) when a manufacturer ceases to be an intermediate volume manufacturer in 2003 through 2017 due to the aggregation requirements in majority ownership situation.

(B) *Decreases in California Production Volume*. If a manufacturer's average California production volume falls below 4 500, 10 000, or 60 000 units of new PCs, LDTs, and MDVs, based on the average number of vehicles produced and delivered for sale for the three previous consecutive model years, the manufacturer shall be treated as a small volume, independent low volume, or intermediate volume manufacturer, as applicable, and shall be subject to the requirements for a small volume, independent low volume, or intermediate volume manufacturer beginning with the next model year.

(C) *Calculating California Production Volume in Change of Ownership Situations*. Where a manufacturer experiences a change in ownership in a particular model year, the change will affect application of the aggregation requirements on the manufacturer starting with the next model year. When a manufacturer is simultaneously producing two model years of vehicles at the time of a change of ownership, the basis of determining next model year must be the earlier model year. The manufacturer's small, independent low, or intermediate volume manufacturer status for the next model year shall be based on the average California production volume in the three previous consecutive model years of those manufacturers whose production volumes must be aggregated for that next model year. For example, where a change of ownership during the 2010 calendar year occurs and the manufacturer is producing both 2010 and 2011 model year vehicles resulting in a requirement that the production volume of Manufacturer A be aggregated with the production volume of Manufacturer B, Manufacturer A's status for the 2011 model year will be based on the production volumes of Manufacturers A and B in the 2008 – 2010 model years. Where the production volume of Manufacturer A must be aggregated with the production volumes of Manufacturers B and C for the 2010 model year, and during that model year a change in ownership eliminates the requirement that Manufacturer B's production volume be aggregated with Manufacturer A's, Manufacturer A's status for the 2011 model year will be based on the production volumes of Manufacturers A and C in the 2008 – 2010 model years. In either case, the lead time provisions in subdivisions 1962.1(b)(7)(A) and (B) will apply.

(c) *Partial ZEV Allowance Vehicles* (*PZEVs*).

(1) *Introduction*. This subdivision 1962.1(c) sets forth the criteria for identifying vehicles delivered for sale in California as PZEVs. The PZEV is a vehicle that cannot be certified as a ZEV but qualifies for a PZEV allowance of at least 0.2.

(2) *Baseline PZEV Allowance*. In order for a vehicle to be eligible to receive a PZEV allowance, the manufacturer must demonstrate compliance with all of the following requirements. A qualifying vehicle will receive a baseline PZEV allowance of 0.2.

(A) *SULEV Standards*. For 2009 through 2013 model years, certify the vehicle to the 150 000 – mile SULEV exhaust emission standards for PCs and LDTs in subdivision 1961(a)(1). Bi-fuel, fuel-flexible and dual-fuel vehicles must certify to the applicable 150 000 – mile SULEV exhaust emission standards when operating on both fuels. For 2014 through 2017 model years, certify the vehicle to the 150 000 – mile SULEV 20 or 30 exhaust emission standards for PCs and LDTs in subdivision 1961. 2 (a) (1), or to the 150 000 – mile SULEV exhaust emission standards for PCs and LDTs in subdivision 1961 (a) (1). Bi-fuel, fuel flexible and dual-fuel vehicles must certify to the applicable 150 000 – mile SULEV exhaust emission standards when operating on both fuels;

(**B**) *Evaporative Emissions*. For 2009 through 2013 model years, certify the vehicle to the evaporative emission standards in subdivision 1976(b)(1)(E)(zero-fuel evaporative emissions standards). For 2014 through 2017 model years, certify the vehicle to the evaporative emission standards in subdivision 1976(b)(1)(G) or subdivision 1976 (b) (1)(E);

(**C**) *OBD*. Certify that the vehicle will meet the applicable on-board diagnostic requirements in sections 1968. 1 or 1968. 2, as applicable, for 150 000 miles; and

(**D**) *Extended Warranty*. Extend the performance and defects warranty period set forth in subdivision 2037(b)(2) and 2038(b)(2) to 15 years or 150 000 miles, whichever occurs first except that the time period is to be 10 years for a zero-emission energy storage device used for traction power (such as a battery, ultracapacitor, or other electric storage device).

(3) *Zero-Emission VMT PZEV Allowance*.

(A) *Calculation of Zero-Emission VMT Allowance*. A vehicle that meets the requirements of subdivision 1962. 1 (c) (2) and has zero-emission vehicle miles traveled ("VMT") capability will generate an additional zero-emission VMT PZEV allowance calculated as follows:

Range	Zero-emission VMT Allowance
EAERu < 10 miles	0. 0
EAERu ≥ 10 to 40 miles	EAER u × (1 – UFRcda) /11. 028
EAERu ≥ 40 miles	3. 627 × (1 – UFu) Where: n = 40 × (R_{cda}/EAERu)

A vehicle cannot generate more than 1. 39 zero-emission VMT PZEV allowances.

The urban equivalent all-electric range (EAERu) and urban charge depletion range actual (Rcda) shall be determined in accordance with sections G. 11. 4 and

G. 11. 9, respectively, of the "California Exhaust Emission Standards and Test Procedures for 2009 through 2017 Model Zero-Emission Vehicles and Hybrid Electric Vehicles, in the Passenger Car, Light-Duty Truck and Medium-Duty Vehicle Classes," adopted December 17, 2008, and last amended May 30, 2014, incorporated by reference in section 1962. 1 (h). The utility Factor (UF) shall be determined according to SAE International's Surface Vehicle Information Report J2841 SEP2010 (Revised September 2010), incorporated by reference herein, from the Fleet Utility Factors (FUF) Table in Appendix B or using a polynomial curve fit with "FUF Fit" coefficients from Table 2 Utility Factor Equation Coefficients.

(**B**) ***Alternative Procedures***. As an alternative to determining the zero-emission VMT allowance in accordance with the preceding section 1962. 1 (c)(3)(A), a manufacturer may submit for Executive Officer approval an alternative procedure for determining the zero-emission VMT potential of the vehicle as a percent of total VMT, along with an engineering evaluation that adequately substantiates the zero-emission VMT determination. For example, an alternative procedure may provide that a vehicle with zero-emissions of one regulated pollutant (e. g., NOx) and not another (e. g., NMOG) will qualify for a zero-emission VMT allowance of 1. 5.

(**4**) ***PZEV Allowance for Advanced ZEV Componentry***. A vehicle that meets the requirements of subdivision 1962. 1 (c)(2) may qualify for an advanced componentry PZEV allowance as provided in this section 1962. 1 (c)(4).

(**A**) ***Use of High Pressure Gaseous Fuel or Hydrogen Storage System***. A vehicle equipped with a high pressure gaseous fuel storage system capable of refueling at 3600 pounds per square inch or more and operating exclusively on this gaseous fuel shall qualify for an advanced componentry PZEV allowance of 0. 2. A vehicle capable of operating exclusively on hydrogen stored in a high pressure system capable of refueling at 5 000 pounds per square inch or more, stored in nongaseous form, or at cryogenic temperatures, shall instead qualify for an advanced componentry PZEV allowance of 0. 3.

(**B**) ***Use of a Qualifying HEV Electric Drive System***.

1. ***Classification of HEVs***. HEVs qualifying for additional advanced componentry PZEV allowance or allowances that may be used in the AT PZEV category are classified in one of four types of HEVs based on the criteria in the following table.

Characteristics	*Type D*	*Type E*	*Type F*	*Type G*
Electric Drive System Peak Power Output	≥10 kW	≥ 50 kW	Zero-Emission VMT allowance; ≥ 10 mile all-electric UDDS range	Zero-Emission VMT allowance; ≥ 10 mile all-electric US06 range
Traction Drive System Voltage	≥ 60 Volts	≥ 60 volts	≥ 60 volts	≥ 60 volts
Traction Drive Boost	Yes	Yes	Yes	Yes
Regenerative Braking	Yes	Yes	Yes	Yes
Idle Start/Stop	Yes	Yes	Yes	Yes

2. [*Reserved*]

3. [*Reserved*]

4. [*Reserved*]

5. *Type D HEVs.* A PZEV that the manufacturer demonstrates to the reasonable satisfaction of the Executive Officer meets all of the criteria for a Type D HEV qualifies for an additional advanced componentry allowance of 0. 4 in the 2009 through 2011 model years, 0. 35 in the 2012 through 2014 model years, and 0. 25 in the 2015 through 2017 model years.

6. *Type E HEVs.* A PZEV that the manufacturer demonstrates to the reasonable satisfaction of the Executive Officer meets all of the criteria for a Type E HEV qualifies for an additional advanced componentry allowance of 0. 5 in the 2009 through 2011 model years, 0. 45 in the 2012 through 2014 model years, and 0. 35 in the 2015 through 2017 model years.

7. *Type F HEVs.* A PZEV that the manufacturer demonstrates to the reasonable satisfaction of the Executive Officer meets all of the criteria for a Type F HEV, including a-chieving 10 miles or more of all-electric UDDS range, qualifies for an additional advanced componentry allowance of 0. 72 in the 2009 through 2011 model years, 0. 67 in the 2012 through 2014 model years, and 0. 57 in the 2015 through 2017 model years.

8. *Type G HEVs.* A PZEV that the manufacturer demonstrates to the reasonable satisfaction of the Executive Officer meets all of the criteria for a Type G HEV, including a-chieving 10 miles or more of all-electric US06 range, qualifies for an additional advanced componentry allowance of 0. 95 in the 2009 through 2011 model years, 0. 9 in the 2012 through 2014 model years, and 0. 8 in the 2015 through 2017 model years.

9. *Severability*. In the event that all or part of subdivision 1962. 1（c）（4）（B） 1. - 8. is found invalid, the remainder of section 1962. 1 remains in full force and effect.

（5） *PZEV Allowance for Low Fuel-Cycle Emissions*. A vehicle that makes exclusive use of fuel（s）with very low fuel-cycle emissions shall receive a PZEV allowance of 0. 3. In order to receive the PZEV low fuel-cycle emissions allowance, a manufacturer must demonstrate to the Executive Officer, using peer-reviewed studies or other relevant information, that NMOG emissions associated with the fuel（s）used by the vehicle （on a grams/ mile basis） are lower than or equal to 0. 01 grams/mile. Fuel-cycle emissions must be calculated based on near-term production methods and infrastructure assumptions, and the uncertainty in the results must be quantified.

（6） *Calculation of PZEV Allowance*.

（A） *Calculation of Combined PZEV Allowance for a Vehicle*. The combined PZEV allowance for a qualifying vehicle in a particular model year is the sum of the PZEV allowances listed in this subdivision 1962. 1（c）（6）, multiplied by any PZEV introduction phase-in multiplier listed in subdivision 1962. 1（c）（7）, subject to the caps in subdivision 1962. 1（c）（6）（B）.

1. *Baseline PZEV Allowance*. The baseline PZEV allowance of 0. 2 for vehicles meeting the criteria in subdivision 1962. 1（c）（2）;

2. *Zero-Emission VMT PZEV Allowance*. The zero-emission VMT PZEV allowance, if any, determined in accordance with subdivision 1962. 1（c）（3）;

3. *Advanced Componentry PZEV Allowance*. The advanced ZEV componentry PZEV allowance, if any, determined in accordance with subdivision 1962. 1（c）（4）; and

4. *Fuel-Cycle Emissions PZEV Allowance*. The fuel-cycle emissions PZEV allowance, if any, determined in accordance with subdivision 1962. 1（c）（5）.

（B） *Caps on the Value of an AT PZEV Allowance*.

1. *Cap for 2009 through 2017 Model Year Vehicles*. The maximum value an AT PZEV may earn before phase-in multipliers, including the baseline PZEV allowance, is 3. 0.

2. ［*Reserved*］.

（7） *PZEV Multipliers*.

（A） ［*Reserved*］.

（B） *Introduction Phase-In Multiplier for PZEVs That Earn a Zero-Emission VMT Allowance*. Each 2009 through 2011 model year PZEV that earns a zero-emission

VMT allowance under section 1962. 1 (c) (3) and is sold to a California motorist or is leased for three or more years to a California motorist who is given the option to purchase or re-lease the vehicle for two years or more at the end of the first lease term, qualifies for a phase-in multiplier of 1. 25. This subdivision 1962. 1(c)(7)(B) multiplier will no longer be available after model year 2011.

(d) *Qualification for ZEV Multipliers and Credits*.

(1) [*Reserved*] .

(2) [*Reserved*] .

(3) [*Reserved*] .

(4) [*Reserved*] .

(5) *Credits for 2009 through 2017 Model Year ZEVs*.

(A) *ZEV Tiers for Credit Calculations*. Credits from a particular ZEV are based on the assignment of a given ZEV into one of the following eight ZEV tiers:

ZEV Tier	UDDS ZEV Range (miles)	Fast Refueling Capability
NEV	No minimum	N/A
Type 0	< 50	N/A
Type I	≥ 50 , <75	N/A
Type I. 5	≥ 75 , <100	N/A
Type II	≥ 100	N/A
Type III	≥ 100	Must be capable of replacing 95 miles (UDDS ZEV range) in ≤ 10minutes per section 1962. 1(d)(5)(B)
	≥ 200	N/A
Type IV	≥ 200	Must be capable of replacing 190 miles (UDDS ZEV range) in ≤15 minutes per section 1962. 1(d)(5)(B)
Type V	≥ 300	Must be capable of replacing 285 miles (UDDS ZEV range) in ≤ 15 minutes per section 1962. 1(d)(5)(B)

Type I. 5x and Type IIx vehicles are defined in subdivision 1962. 1 (d)(5)(G) and (i)(10).

(B) Fast Refueling. The "fast refueling capability" requirement for a 2009 through 2017 model year Type III, IV, or V ZEV in subdivision 1963(d)(5)(A) will be considered met if the Type III ZEV has the capability to accumulate at least 95 miles of UDDS range in 10 minutes or less and the Type IV or V ZEVs that utilize more than one ZEV fuel, such as plug – in fuel cell vehicles, the Executive Officer may choose to waive these subdivision 1962. 1(d) (5) (B) fast refueling requirements and base the amount of credit earned on UDDS ZEV range, as specified in subdivision 1962. 1(d) (5) (A).

(C) *Credits for 2009 through 2017 Model Year ZEVs.* A 2009 through 2017 model-year ZEV, including a Type I. 5x and Type IIx, other than a NEV or Type 0, earns 1 ZEV credit when it is produced and delivered for sale in California. A 2009 through 2017 model-year ZEV earns additional credits based on the earliest year in which the ZEV is placed in service in California (not earlier than the ZEV's model year). The vehicle must be delivered for sale and placed in service in a Section 177 state or in California in order to earn the total credit amount. The total credit amount will be earned in the state (i. e. California or a Section 177 state) in which the vehicle was delivered for sale. The following table identifies the total credits that a ZEV in each of the eight ZEV tiers will earn, including the credit not contingent on placement in service, if it is placed in service in the specified calendar year or by June 30 after the end of the specified calendar year. A vehicle is not eligible to receive credits if it is placed in service after December 31, five calendar years after the model year. For example, if a vehicle is produced in 2012, but does not get placed until January 1, 2018, the vehicle would no longer be eligible for ZEV credits.

Total Credit Earned by ZEV Type and Model Year for Production and Delivery for Sale and for Placement

Tier	Calendar Year in Which ZEV is Placed in Service	
	2009 – 2011	2012 – 2017
NEV	0. 30	0. 30
Type 0	1	1
Type I	2	2
Type I. 5	2. 5	2. 5
Type I. 5x	n/a	2. 5
Type II	3	3
Type IIx	n/a	3
Type III	4	4
Type IV	5	5
Type V	7	2012 – 2014: 7 2015 – 2017: 9

* As specified in subdivision 1962. 1(d) (5) (B)

(D) *Multiplier for Certain ZEVs.* 2009 through 2011 model-year ZEVs, excluding NEVs or Type 0 ZEVs, shall qualify for a multiplier of 1. 25 if either sold to a motorist or leased for three or more years to a motorist who is given the option to purchase or re-lease the vehicle for two years or more at the end of the first lease term. This subdivision 1962. 1 (d) (5) (D) multiplier will no longer be available after model year 2011.

(E) *Counting Specified ZEVs Placed in a Section 177 State and in California.*

1. Provisions for 2009 Model Year.

a. Large volume manufacturers and intermediate volume manufacturers with credits earned from ZEVs, excluding NEVs and Type 0 ZEVs, that are either certified to the California ZEV standards or approved as part of an advanced technology demonstration program and are placed in service in a section 177 state, may be counted towards compliance with the California percentage ZEV requirements in subdivision 1962. 1 (b) , including the requirements in subdivision 1962. 1 (b) (2) (B) , as if they were delivered for sale and placed in service in California.

b. Large volume manufacturers and intermediate volume manufacturers with credits earned from ZEVs, excluding NEVs and Type 0 ZEVs, that are certified to the California ZEV standards or approved as part of an advanced technology demonstration program and are placed in service in California may be counted towards the percentage ZEV requirements of all section 177 states, including requirements based on subdivision 1962. 1 (b) (2) (B) .

2. *Provisions for 2010 through 2017 Model Years.* Large volume manufacturers and intermediate volume manufacturers with credits earned from ZEVs, including Type I. 5x and Type IIx vehicles, and excluding NEVs and Type 0 ZEVs, that are either certified to the California ZEV standards applicable for the ZEV's model year or approved as part of an advanced technology demonstration program and are placed in service in California or in a section 177 state may be counted towards compliance in California and in all section 177 states, with the percentage ZEV requirements in subdivision 1962. 1 (b) , provided that the credits are multiplied by the ratio of a manufacturer's applicable production volume for a model year, as specified in subdivision 1962. 1 (b) (1) (B) , in the state receiving credit to the manufacturer's applicable production volume (hereafter, "proportional value") , as specified in section 1962. 1 (b) (1) (B) , for the same model year in California. Credits generated in a section 177 state will be earned at the proportional value in the section 177 state, and earned in California at the full value specified in subdivision 1962. 1 (d) (5) (C) . However, credits generated by 2010 and 2011 model-year vehicles produced, delivered for sale, and placed in service or as part of an advanced technology demonstration program in California to meet any section 177 state's requirements that implement subdivision 1962. 1(b)(2)(B) are exempt from proportional value, with the number of credits exempted from proportional value allowed being limited to the number of credits needed to satisfy a manufacturer's section 177 state's requirements that implement subdivision 1962. 1(b)(2)(B) 1. b. The table below specifies the qualifying model years for each ZEV type that may be counted towards compliance in all section 177 states.

Vehicle Type	Model Years:
Type I, I. 5, or II ZEV	2009 – 2017
Type III, IV, or V ZEV	2009 – 2017
Type I. 5x or Type IIx	2012 – 2017

3. *Optional Section 177 State Compliance Path*. Large volume manufacturers and intermediate volume manufacturers that choose to elect the optional Section 177 state compliance path must notify the Executive Officer and each Section 177 state in writing no later than September 1, 2014.

a. *Additional 2016 and 2017 Model Year ZEV Requirements*. Large volume manufacturers and intermediate volume manufacturers that elect the optional Section 177 state compliance path must generate additional 2012 through 2017 model year ZEV credits, including no more than 50% Type 1. 5x and Type IIx vehicle credits and excluding all NEV, Type 0 ZEV credits, and transportation system credits, in each Section 177 state to fulfill the following percentage requirements of their sales volume determined under subdivision 1962. 1(b)(1)(B):

Model Years	Additional Section 177 State ZEV Requirements
2016	0. 75%
2017	1. 50%

Subdivision 1962. 1(d)(5)(E) 2. shall not apply to any ZEV credits used to meet a manufacturer's additional 2016 and 2017 model year ZEV requirements under this subdivision 1962. 1(d)(5)(E) 3. a. ZEVs produced to meet a manufacturer's additional

2016 and 2017 model year ZEV requirements under this subdivision 1962. 1 (d) (5) (E) 3. a. must be placed in service in the Section 177 states no later than June 30, 2018.

i. *Trading and Transferring ZEV Credits within the West Region Pool and East Region Pool*. Starting in model year 2016, manufacturers may trade or transfer 2012 through 2017 model year ZEV credits, used to meet the requirements in subdivisions 1962. 1 (d) (5) (E) 3. a. and c. , within the West Region pool, and will incur no premium on their credit values. For example, for a manufacturer to make up a 2016 model year shortfall of 100 credits in State X, the manufacturer may transfer 100 (2016 model year) ZEV credits from State Y, within the West Region pool. Manufacturers may trade or transfer specific model year ZEV credits, used to meet the same model year requirements in subdivisions 1962. 1 (d) (5) (E) 3. a. and c. , within the East Region pool, and will incur no premium on their credit values. For example, for a manufacturer to make up a 2016 model year shortfall of 100 credits in State W, the manufacturer may transfer 100 (2016 model year) ZEV credits from State Z, within the East Region pool.

ii. *Trading and Transferring ZEV Credits between the West Region Pool and East Region Pool*. Manufacturers may trade or transfer specified model year ZEV credits used to meet the same model yaer. requirements in subdivision 1962. 1 (d) (5) (E) 3. c. between the West Region pool and the East Region pool; however, any credits traded or transferred will incur a premium of 30% of their value. For example, in order for a manufacturer to make up a 2016 model year shortfall of 100 credits in the West Region Pool, the manufacturer may transfer 130 (2016 model year) ZEV credits from the East Region Pool. No credits may be traded or transferred to the East Region pool or West Region pool from a manufacturer's California ZEV bank, or from the East Region pool or West Region pool to a manufacturer's California ZEV bank.

b. *Reduced TZEV Percentages*. Large volume manufacturers and intermediate volume manufacturers that elect the optional Section 177 state compliance path and that fully comply with the additional 2016 and 2017 model year ZEV requirements in this subdivision 1962. 1 (d) (5) (E) 3. a. are allowed to meet TZEV percentages reduced from the allowed TZEV percentages in subdivision 1962. 1 (b) (2) (D) 2. and 3. in 2015 through 2017 model year in each Section 177 state as enumerated below:

Model Year	*2015*	*2016*	*2017*
Existing TZEV Percentage	3.00%	3.00%	3.00%
Section 177 State Adjustment for Optional Compliance Path for TZEVs	75.00%	80.00%	85.00%
New Section 177 State Optional Compliance Path TZEV Percentage	2.25%	2.40%	2.55%

Manufacturers may meet the reduced TZEV percentages above with credits from ZEVs or credits from TZEVs. These reduced TZEV percentages also reduce the total ZEV percent requirement, as illustrated in subdivision 1962.1(d)(5)(E) 3. c.

i. *Trading and Transferring TZEV Credits within the West Region Pool and the East Region Pool*. Starting in model year 2015, manufacturers may trade or transfer specified TZEV credits used to meet the same model year subdivision 1962.1(d)(5)(E) 3. c. requirements within the West Region pool, and will incur no premium on their credit values. For example, for a manufacturer to make up a 2016 shortfall of 100 credits in State X, the manufacturer may transfer 100 (2016 model year) TZEV credits from State Y, within the West Region pool. Manufacturers may trade or transfer TZEV credits to meet the same model year. subdivision 1962.1(d)(5)(E) 3. c. within the East Region pool, and will incur no premium on their credit values. For example, for a manufacturer to make up a 2016 model year shortfall of 100 credits in State W, the manufacturer may transfer 100 (2016 model year) TZEV credits from State Z, within the East Region pool.

ii. *Trading and Transferring TZEV Credits between the West Region Pool and the East Region Pool*. Manufacturers may trade or transfer specified TZEV credits used to meet the the same model year percentages in subdivision 1962.1(d)(5)(E) 3. c. between the West Region pool and the East Region pool; however, any credits traded or transferred will incur a premium of 30% of their value. For example, in order for a manufacturer to make up a 2016 model year shortfall of 100 credits in the West Region Pool, the manufacturer may transfer 130(2016 model year) TZEV credits from the East Region Pool. No credits may be traded or transferred to the East Region pool or West Region pool from a manufacturer's California ZEV bank, or from the East Region pool or West Region pool to a manufacturer's California ZEV bank.

c. *Total Requirement Percentages*. Requirements for the minimum ZEV floor, and allowed percentages for AT PZEVs and PZEVs in subdivision 1962.1(b) remain in effect for large and intermediate volume manufacturers choosing the optional section 177 state compliance path in each section 177 state. However, the optional Section 177 compliance path requires manufacturers to meet additional ZEV requirements and allows manufacturers to meet reduced TZEV percentages as described above in subdivision 1962.1(d)(5)(E) 3. a. and b. The table below enumerates the total annual percentage obligation in each section 177 state for the 2015 through 2017 model years if the manufacturer elects the optional section 177 state compliance path and produces the minimum number of credits required to meet its minimum ZEV floor and the maximum percentage allowed to be met with credits from TZEVs, AT PZEVs and PZEVs.

Years	*Total ZEV Percent Requirement for Optional Compliance Path*	*Minimum ZEV Floor for Optional Compliance Path*	*TZEVs for Optional Compliance Path*	*AT PZEVs (no change)*	*PZEVs (no change)*
2015	13. 25%	3. 00%	2. 25%	2. 00%	6. 00%
2016	14. 15%	3. 75%	2. 40%	2. 00%	6. 00%
2017	15. 05%	4. 50%	2. 55%	2. 00%	6. 00%

d. Reporting Requirements. On an annual basis, by May 1st of the calendar year following the close of a model year, each manufacturer that elects the optional section 177 state compliance path shall submit, in writing, to the Executive Officer and each section 177 state a report, including an itemized list, that demonstrates the manufacturer has met the requirements of this subdivision 1962.1(d)(5)(E) 3. in each section 177 state as well as in the East Region pool and in the West Region pool. The itemized list shall include the following:

i. The manufacturer's total applicable volume of PCs and LDTs delivered for sale in each section 177 state within the pool, as determined under subdivision 1962.1(b)(1)(B).

ii. Make, model, vehicle identification number, credit earned, and section 177 state where delivery for sale and placement in service for ZEV occurred to meet the manufacturer's additional ZEV obligation under subdivision 1962.1(d)(5)(E) 3. a.

iii. Make, model, credit earned, and section 177 state where delivery for sale of TZEVs occurred and section 177 state where delivery for sale and placement in service of each ZEV occurred to meet manufacturer's requirements under subdivision 1962.1(d)(5)(E) 3. c.

e. Failure to Meet Optional Section 177 State Compliance Path Requirements.

A manufacturer that elects the optional section 177 state compliance path and does not meet the requirements in subdivision 1962. 1 （d）（5）（E） 3. a. by June 30, 2018 in all section 177 states within an applicable pool shall be treated as subject to the total ZEV percentage requirements in section 1962. 1 （b） for all future model years in each section 177 state and the pooling provisions in subdivision 1962. 1 （d）（5）（E） 3. a. shall not apply. Any transfers of ZEV credits between section 177 states will be null and void, and ZEV credits will return to the section 177 state in which the credits were earned. A manufacturer that e-lects the optional section 177 state compliance path and does not meet the percentages in subdivision 1962. 1（d）（5）（E） 3. b. in a model year or make up their deficit within the specified time and with the specified credits allowed by subdivision 1962. 1 （g）（7）（A） in all section 177 states within an applicable pool shall be treated as subject to the total ZEV percentage requirements in section 1962. 1 （b） for the 2015 through 2017 model years and the pooling provisions in subdivision 1962. 1 （d）（5）（E） 3. b. shall not apply. Any transfers of TZEV credits between section 177 states will be null and void if a manufacturer fails to comply, and TZEV credits nill return to the section 177 state in which the credits nere earned. . Penalties shall be calculated separately by each section 177 state where a manufacturer fails to make up the ZEV deficits by the end of the 2018 model year.

f. The provisions in section 1962. 1 shall apply to a manufacturer electing the op-tional Section 177 state compliance path, except as specifically modified by this subdivision 1962. 1（d）（5）（E） 3.

（**F**） *NEVs*. Beginning in 2010 model year, to be eligible for the credit amount in subdivision 1962. 1（d）（5）（C）, NEVs must meet the following specifications and re-quirements in this subdivision 1962. 1（d）（5）（F）:

1. *Specifications*. A 2010 through 2017 model year NEV earns credit when it meets all the following specifications:

a. *Acceleration*. The vehicle has a 0 – 20 mph acceleration of 6. 0 seconds or less when operating with a payload of 332 pounds and starting with the battery at a 50% state of charge.

b. *Top Speed*. The vehicle has a minimum top speed of 20 mph when operating with a payload of 332 pounds and starting with the battery at a 50% state of charge. The vehicle's top speed shall not exceed 25 mph when tested in accordance with 49 CFR 571. 500 （68 FR 43972, July 25, 2003）.

c. *Constant Speed Range*. The vehicle has a minimum 25 – mile range when operating at constant top speed with a payload of 332 pounds and starting with the battery at 100% state of charge.

2. *Battery Requirement*. A 2010 through 2017 model year NEV must be e-quipped with one or more sealed, maintenance-free batteries.

3. *Warranty Requirement*. A 2010 through 2017 model year NEV drive train, including battery packs, must be covered for a period of at least 24 months. The first 6 months of the NEV warranty period must be covered by a full warranty; the remaining warranty period may be optional extended warranties (available for purchase) and may be prorated. If the extended warranty is prorated, the percentage of the battery pack's original value to be covered or refunded must be at least as high as the percentage of the prorated coverage period still remaining. For the purpose of this computation, the age of the battery pack must be expressed in intervals no larger than three months. Alternatively, a manufacturer may cover 50 percent of the original value of the battery pack for the full period of the extended warranty.

4. Prior to allowance approval, the Executive Officer may request that the manufacturer provide copies of representative vehicle and battery warranties.

5. *NEV Charging Requirements*. Model year 2014 through 2017 NEVs must meet charging connection standard portion of the requirements specified in subdivision 1962.3(c)(2).

(G) *Type I.5x and Type IIx Vehicles*. Beginning in 2012 model year, to be eligible for the credit amount in subdivision 1962.1(d)(5)(C), Type I.5x and Type IIx vehicles must meet the following specifications and requirements:

1. *PZEV Requirements*. Type I.5x and Type IIx vehicles must meet all PZEV requirements, specified in subdivision 1962.1(c)(2)(A) through (D).

2. *Type G Requirements*. Type I.5x and Type IIx vehicles must meet the requirements for Type G advanced componentry allowance, specified in subdivision 1962.1 (c)(4)(B).

3. *APU Operation*. The vehicle's UDDS range after the APU first starts and enters "charge sustaining hybrid operation" must be less than or equal to the vehicle's UDDS all-electric test range prior to APU start. The vehicle's APU cannot start under any user-selectable driving mode unless the energy storage system used for traction power is fully depleted.

4. *Minimum Zero Emission Range Requirements*.

Vehicle Category	Zero Emission UDDS Range
Type I. 5x	≥ 75 miles, < 100 miles
Type IIx	≥ 100 miles

(e) [Reserved].

(f) *Extended Service Multiplier for 1997 – 2003 Model Year ZEVs and PZEVs With ≥ 10 Mile Zero-Emission Range*. Except in the case of a NEV, an additional ZEV or PZEV multiplier will be earned by the manufacturer of a 1997 through 2003 model year ZEV, or PZEV with ≥ 10 mile zero-emission range for each full year it is registered for operation on public roads in California beyond its first three years of service, in the 2009 through 2011 calendar years. For additional years of service starting earlier than April 24, 2003, the manufacturer will receive 0.1 times the ZEV credit that would be earned by the vehicle if it were leased or sold new in that year, including multipliers, on a year-by-year basis beginning in the fourth year after the vehicle is initially placed in service. For additional years of service starting April 24, 2003 or later, the manufacturer will receive 0.2 times the ZEV credit that would be earned by the vehicle if it were leased or sold new in that year, including multipliers, on a year-by-year basis beginning in the fourth year after the vehicle is initially placed in service. The extended service multiplier is reported and earned in the year following each continuous year of service. Additional credit cannot be earned after model year 2011.

(g) *Generation and Use of Credits; Calculation of Penalties*

(1) *Introduction*. A manufacturer that produces and delivers for sale in California ZEVs or PZEVs in a given model year exceeding the manufacturer's ZEV requirement set forth in subdivision 1962.1(b) shall earn credits in accordance with this subdivision 1962.1(g).

(2) *Credit Calculations*.

(A) *Credits from ZEVs*. For model years 2009 through 2014, the amount of g/mi credits earned by a manufacturer in a given model year from ZEVs shall be expressed in units of g/mi NMOG, and shall be equal to the number of credits from ZEVs produced and delivered for sale in California that the manufacturer applies towards meeting the ZEV requirements for the model year subtracted from the number of ZEVs produced and delivered for sale in California by the manufacturer in the model year and then multiplied by the NMOG fleet average requirement for PCs and LDT1s, or LDT2s as applicable, for 2009 through 2011 model years, and for PCs and LDT1s for 2012 through 2014 model years.

For model years 2015 through 2017, the amount of credits earned by a manufacturer in a given model year from ZEVs shall be expressed in units of credits and shall be equal to the number of credits from ZEVs produced and delivered for sale in California that the manufacturer applies towards meeting the ZEV requirements, or, if applicable, requirements specified under subdivision 1962. 1 (d) (5) (E) 3. , for the model year subtracted from the number of ZEV credits produced and delivered for sale in California by the manufacturer in the model year or model years.

(B) *Credits from PZEVs*. For model years 2009 through 2014, the amount of g/mi credits from PZEVs earned by a manufacturer in a given model year shall be expressed in units of g/mi NMOG, and shall be equal to the total number of PZEVs produced and delivered for sale in California that the manufacturer applies towards meeting its ZEV requirement for the model year subtracted from the total number of PZEV allowances from PZEVs produced and delivered for sale in California by the manufacturer in the model year and then multiplied by the NMOG fleet average requirement for PCs and LDT1s, or LDT2s as applicable, for 2009 through 2011 model years, and for PCs and LDT1s for 2012 through 2014 model years.

For model years 2015 through 2017, the amount of credits earned by a manufacturer in a given model year from PZEVs shall be expressed in units of credits, and shall be equal to the number of credits from PZEVs produced and delivered for sale in California that the manufacturer applies towards meeting the ZEV requirements, or, if applicable, requirements specified under subdivision 1962. 1 (d) (5) (E) 3. , for the model year subtracted from the number of PZEV credits produced and delivered for sale in California by the manufacturer in the model year or model years.

(C) *Separate Credit Accounts*. The number of credits from a manufacturer's [i] ZEVs, [ii] Type I. 5x and Type IIx vehicles, [iii] TZEVs, [iv] AT PZEVs, [v] all other PZEVs, and [vi] NEVs shall each be maintained separately.

(D) *Rounding Credits*. For model year 2012 through 2014, ZEV credits and debits shall be rounded to the nearest $1/1\,000^{th}$ only on the final credit and debit totals using the conventional rounding method. For model year 2015 through 2017, ZEV credits and debits shall be rounded to the nearest $1/100^{th}$ only on the final credit and debit totals using the conventional rounding method.

(E) *Converting g/mi NMOG ZEV Credits to ZEV Credits*. After model year

2014 compliance, all manufacturer ZEV, Type I. 5x and Type IIx, TZEV, AT PZEV, PZEV, and NEV accounts will be converted from g/mi NMOG to credits. Each g/mi NMOG account balance will be divided by 0.035. Starting in model year 2015, credits will no longer be expressed in terms of g/mi credits, but only as credits.

(F) *Converting PZEV and AT PZEV Credits after Model Year 2017*. After model year 2017 compliance, a manufacturer's PZEV and AT PZEV credit accounts will be converted to be used for compliance with requirements specified in subdivision 1962. 2(b). For LVMs, PZEV accounts will be discounted 93. 25%, and AT PZEV accounts will be discounted 75%. For IVMs, PZEV accounts and AT PZEV accounts will be discounted 75%. This will be a one time calculation after model year 2017 compliance is complete.

(3) *ZEV Credits for MDVs and LDTs Other Than LDT1s*. ZEVs and PZEVs classified as MDVs or as LDTs other than LDT1s may be counted toward the ZEV requirement for PCs, LDT1s and LDT2s as applicable, and included in the calculation of ZEV credits as specified in this subdivision 1962. 1(g) if the manufacturer so designates.

(4) *ZEV Credits for Advanced Technology Demonstration Programs*.

(A) *TZEVs*. For 2009 through 2014 model years, TZEVs placed in a California advanced technology demonstration program for a period of two or more years, may earn ZEV credits even if it is not "delivered for sale" or registered with the California Department of Motor Vehicles (DMV). To earn such credits, the manufacturer must demonstrate to the reasonable satisfaction of the Executive Officer that the vehicles will be regularly used in applications appropriate to evaluate issues related to safety, infrastructure, fuel specifications or public education, and that for 50 percent or more of the first two years of placement the vehicle will be operated in California. Such a vehicle is eligible to receive the same allowances and credits that it would have earned if placed in service. To determine vehicle credit, the model year designation for a demonstration vehicle shall be consistent with the model year designation for conventional vehicles placed in the same timeframe. Manufacturers may earn credit for as many as 25 vehicles per model, per ZEV state, per year under this subdivision 1962. 1(g)(4). A manufacturer's vehicles in excess of the 25 – vehicle cap will not be eligible for advanced technology demonstration program credits.

(B) *ZEVs*. In model years 2009 through 2017, ZEVs, including Type I. 5x and IIx vehicles, excluding NEVs and Type 0 ZEVs, placed in a California advanced technology demonstration program for a period of two or more years, may earn ZEV credits even if it is not "delivered for sale" or registered with the California DMV. To earn such credits, the manufacturer must demonstrate to the reasonable satisfaction of the Executive Officer that the vehicles will be regularly used in applications appropriate to evaluate issues related to safety, infrastructure, fuel specifications or public education, and that for 50 percent or more of the first two years of placement the vehicle will be operated in California. Such a vehicle is eligible to receive the same allowances and credits that it would have earned if placed in service. To determine vehicle credit, the model year designation for a demonstration vehicle shall be consistent with the model year designation for conventional vehicles placed in the same timeframe. Manufacturers may earn credit for as many as 25 vehicles per model, per ZEV state, per year under this subdivision 1962.1(g)(4). A manufacturer's vehicles in excess of the 25 - vehicle cap will not be eligible for advanced technology demonstration program credits.

(5) **ZEV Credits for Transportation Systems**.

(A) **General**. In model years 2009 through 2011, a ZEV placed, for two or more years, as part of a transportation system may earn additional ZEV credits, which may be used in the same manner as other credits earned by vehicles of that category, except as provided in subdivision(g)(5)(C) below. In model years 2012 through 2017, a ZEV, Type I.5x and Type IIx vehicles, or TZEV placed, for two or more years, as part of a transportation system may earn additional ZEV credits, which may be used in the same manner as other credits earned by vehicles of that category, except as provided in subdivision(d)(5)(E) 2. and as provided in subdivision(g)(5)(C) below. In model years 2009 through 2011, an AT PZEV or PZEV placed as part of a transportation system may earn additional ZEV credits, which may be used in the same manner as other credits earned by vehicles of that category, except as provided in subdivision(g)(5)(C) below. A NEV is not eligible to earn credit for transportation systems. To earn such credits, the manufacturer must demonstrate to the reasonable satisfaction of the Executive Officer that the vehicle will be used as a part of a project that uses an innovative transportation system as described in subdivision(g)(5)(B) below.

(B) **Credits Earned**. In order to earn additional credit under this section (g)(5), a project must at a minimum demonstrate [i] shared use of ZEVs, Type I.5x and Type IIx vehicles, TZEVs, AT PZEVs or PZEVs, and [ii] the application of "intelligent" new technologies such as reservation management, card systems, depot management, location management, charge billing and real-time wireless information systems. If, in addition to factors [i] and [ii] above, a project also features linkage to transit, the project may receive further additional credit. For ZEVs only, not including NEVs, a project that features linkage to transit, such as dedicated parking and charging facilities at transit stations, but does not demonstrate shared use or the application of intelligent new technologies, may also receive additional credit for linkage to transit. The maximum credit awarded per vehicle shall be determined by the Executive Officer, based upon an application submitted by the manufacturer and, if appropriate, the project manager. The maximum credit awarded shall not exceed the following:

Type of Vehicle	Model Year	Shared Use, Intelligence	Linkage to Transit
PZEV	through 2011	2	1
AT PZEV	through 2011	4	2
TZEV	2009 through 2011	4	2
ZEV	2009 through 2011	6	3
TZEV	2012 through 2017	0.5	0.5
ZEV and Type I.5x and Type IIx vehicles	2012 through 2017	0.75	0.75

(C) Cap on Use of Transportation System Credits.

1. ZEVs. Credits earned or allocated by ZEVs or Type I.5x and Type IIx vehicles pursuant to this subdivision (g)(5), not including all credits earned by the vehicle itself, may be used to satisfy up to one-tenth of a manufacturer's ZEV obligation in any given model year, and may be used to satisfy up to one-tenth of a manufacturer's ZEV obligation which must be met with ZEVs, as specified in subdivision 1962.1 (b)(2)(D) 3.

2. TLEVs. Credits earned or allocated by TZEVs pursuant to this subdivision (g)(5), not including all credits earned by the vehicle itself, may be used to satisfy up to one-tenth of a manufacturer's ZEV obligation in any given model year, or, if applicable, up to one-tenth of the total ZEV percentages specified under subdivision 1962.1 (d)(5)(E) 3., but may only be used in the same manner as other credits earned by vehicles of that category.

3. AT PZEVs. Credits earned or allocated by AT PZEVs pursuant to this subdivision (g)(5), not including all credits earned by the vehicle itself, may be used to satisfy up to one-twentieth of a manufacturer's ZEV obligation in any given model year, but may only be used in the same manner as other credits earned by vehicles of that category.

4. PZEVs. Credits earned or allocated by PZEVs pursuant to this subdivision (g)(5), not including all credits earned by the vehicle itself, may be used to satisfy up to one-fiftieth of the manufacturer's ZEV obligation in any given model year, but may only be used in the same manner as other credits earned by vehicles of that category.

(**D**) *Allocation of Transportation System Credits*. Credits shall be assigned by the Executive Officer to the project manager or, in the absence of a separate project manager, to the vehicle manufacturers upon demonstration that a vehicle has been placed in a project for the time specified in subdivision 1962. 1 (g) (5) (A). Credits shall be allocated to vehicle manufacturers by the Executive Officer in accordance with a recommendation submitted in writing by the project manager and signed by all manufacturers participating in the project, and need not be allocated in direct proportion to the number of vehicles placed. Credits will no longer be allocated for vehicles placed in transportation systems after 2017 model year.

(**6**) *Use of ZEV Credits*. For model years 2009 through 2014, a manufacturer may meet the ZEV requirements in any given model year by submitting to the Executive Officer a commensurate amount of g/mi ZEV credits, consistent with subdivision 1962. 1 (b). For model years 2015 through 2017, a manufacturer may meet the ZEV requirements in any given model year by submitting to the Executive Officer a commensurate amount of ZEV credits, consistent with subdivision 1962. 1 (b). Credits in each of the categories may be used to meet the requirement for that category as well as the requirements for lesser credit earning ZEV categories, but shall not be used to meet the requirement for a greater credit earning ZEV category. For example, credits produced from TZEVs may be used to comply with AT PZEV requirements, but not with the portion that must be satisfied with ZEVs. These credits may be earned previously by the manufacturer or acquired from another party.

(**A**) *NEVs*. Credits earned from NEVs offered for sale or placed in service in model years 2001 through 2005 cannot be used to satisfy more than the percentage limits described in the following table:

Model Years	ZEV Obligation that:	Percentage limit for NEVs allowed to meet each Obligation[1]
2009 – 2011	Must be met with ZEVs	50%
2009	May be met with AT PZEVs but not PZEVs	75%
2010 – 2011		50%
2009 – 2011	May be met with PZEVs	No Limit

Model Years	ZEV Obligation that:	Percentage limit for NEVs allowed to meet each Obligation¹
2012 – 2017	Must be met with ZEVs	0%
	May be met with TZEVs and AT PZEVs	50%
	May be met with PZEVs	No Limit

* If applicable, obligation in this table means requirements specified under subdivision 1962. 1(d)(5)(E)3.

Additionally, credits earned from NEVs placed in service in model years 2006 through 2017 can be used to meet the percentage limits described in the following table:

Model Years	ZEV Obligation that:	Percentage Limit for NEVs allowed to meet each Obligation¹:
2009 – 2011	May be met through compliance with Primary Requirements	No Limit
	May be met through compliance with Alternative Requirements, and must be met with ZEVs	0%
	May be met through compliance Alternative Requirements, and may be met with AT PZEVs or PZEVs	No Limit
2012 – 2017	Must be met with ZEVs	0%
	May be met with TZEVs, AT PZEVs, or PZEVs	No Limit

¹ If applicable, obligation in this table means requirements specified under subdivision 1962.1 (d)(5)(E)3.

This limitation applies to NEV credits earned by the same manufacturer or earned by another manufacturer and acquired.

(**B**) *Carry forward provisions for LVMs for 2009 –2011 Model Years*. Credits from ZEVs, excluding credits generated from NEVs, generated from excess production in 2009 through 2011 model years, including those acquired from another party, may be carried forward and applied to the ZEV minimum floor requirement specified in subdivisions 1962. 1 (b)(2)(B) 1. b. and (b)(2)(D) for two subsequent model years. Beginning with the third subsequent model year, those earned credits may no longer be used to satisfy the manufacturer's percentage ZEV obligation that may only be satisfied by credits from ZEVs, but may be used to satisfy the manufacturer's percentage ZEV obligation that may be satisfied by credits from TZEVs, AT PZEVs, or PZEVs. For example, ZEV credit earned in 2010 would retain full flexibility through 2012, after which time that credit could only be used as TZEV, AT PZEV, or PZEV credits.

(C) *Carry forward provisions for manufacturers other than LVMs for 2009 – 2011 Model Years.* Credits generated from ZEVs, excluding credits generated from NEVs, from 2009 through 2011 model year production by manufacturers that are not LVMs may be carried forward by the manufacturer producing the credit until the manufacturer becomes subject to the LVM requirements, after the transition period permitted in subdivision 1962. 1 (b)(7)(A). When subject to the LVM requirements, a manufacturer must comply with the provisions of subdivision 1962. 1 (g)(6)(B).

Credits traded by a manufacturer other than a LVM to any other manufacturer, including a LVM, are subject to subdivision 1962. 1 (g)(6)(B), beginning in the model year in which they were produced (e. g. , a 2009 model year credit traded in calendar year 2010 can only be applied towards the portion of the manufacturer's requirement that must be met with ZEVs through model year 2011; beginning in model year 2012, the credit can only be applied to the portion of the manufacturer's requirement that may be met with TZEVs, AT PZEVs, or PZEVs).

(D) *Type I. 5x and Type IIx Vehicles.* Credits earned from Type I. 5x and Type IIx vehicles offered for sale or placed in service may meet up to 50% of the portion of a manufacturer's requirement that must be met with credits from ZEVs.

(7) *Requirement to Make Up a ZEV Deficit.*

(A) *General.* A manufacturer that produces and delivers for sale in California fewer ZEVs than required in a given model year shall make up the deficit by the end of the third model year by submitting to the Executive Officer a commensurate amount of g/mi credits generated by ZEVs, for model year 2009 through 2014, and the commensurate amount of credits generated by ZEVs for model year 2015 through 2017. The amount of credits required to be submitted shall be calculated by [i] adding the number of ZEVs produced and delivered for sale in California by the manufacturer for the model year to the number of ZEV allowances from partial ZEV allowance vehicles produced and delivered for sale in California by the manufacturer for the model year (for a LVM, not to exceed that permitted under subdivision 1962. 1 (b)(2)), [ii] subtracting that total from the number of ZEV credits required to be produced and delivered for sale in California by the manufacturer for the model year, and, for model year 2009 through 2014 compliance, and [iii] multiplying the resulting value by the fleet average requirements for PCs and LDT1s for the model year in which the deficit is incurred. Credits earned by delivery for sale of Type I. 5x and Type IIx vehicles, TZEV, NEV, AT PZEV, and PZEV are not allowed to be used to fulfill a manufacturer's ZEV deficit; only credits from ZEVs may be used to fulfill a manufacturer's ZEV deficit.

(8) *Penalty for Failure to Meet ZEV Requirements*. Any manufacturer that fails to produce and deliver for sale in California the required number of ZEVs and submit an appropriate amount of g/mi credits, for model years 2009 through 2014, and credits for model years 2015 through 2017, and does not make up ZEV deficits within the specified time allowed by subdivision 1962. 1 (g)(7)(A) shall be subject to the Health and Safety Code section 43211 civil penalty applicable to a manufacturer that sells a new motor vehicle that does not meet the applicable emission standards adopted by the state board. The cause of action shall be deemed to accrue when the ZEV deficits are not balanced by the end of the specified time allowed by subdivision 1962. 1 (g)(7)(A). For the purposes of Health and Safety Code section 43 211, the number of vehicles not meeting the state board's standards shall be equal to the manufacturer's credit deficit, rounded to the to the nearest $1/1\,000^{th}$ for model years 2009 through 2014 and rounded to the nearest $1/100^{th}$ for model years 2015 through 2017, calculated according to the following equations, provided that the percentage of a manufacturer's ZEV requirement for a given model year that may be satisfied with PZEV allowance vehicles or credits from such vehicles may not exceed the percentages permitted under subdivision 1962. 1 (b)(2):

For 2009 through 2014 model years:

(No. of credits required to be generated for the model year) − (Amount of credits submitted for compliance for the model year) / (the fleet average requirement for PCs and LDT1s for the model year)

For 2015 through 2017 model years:

(No. of credits required to be generated for the model year) − (Amount of credits submitted for compliance for the model year)

(h) *Test Procedures*.

(1) *Determining Compliance*. The certification requirements and test procedures for

determining compliance with this section 1962. 1 are set forth in "California Exhaust Emission Standards and Test Procedures for 2009 through 2017 Model Zero-Emission Vehicles and Hybrid Electric Vehicles, in the Passenger Car, Light-Duty Truck and Medium-Duty Vehicle Classes," adopted December 17, 2008, and last amended December 6, 2012, which is incorporated herein by reference.

（2）***NEV Compliance***. The test procedures for determining compliance with subdivision 1962. 1 （d）（5）（F）1. are set forth in ETA-NTP002 （revision 3） "Implementation of SAE Standard J1666 May 93: Electric Vehicle Acceleration, Gradeability, and Deceleration Test Procedure" （December 1, 2004）, and ETA-NTP004 （revision 3） "Electric Vehicle Constant Speed Range Tests" （February 1, 2008）, both of which are incorporated by reference herein.

（i）***ZEV-Specific Definitions***. The following definitions apply to this section 1962. 1.

（1） "Advanced technology PZEV" or "AT PZEV" means any PZEV with an allowance greater than 0. 2 before application of the PZEV early introduction phase-in multiplier.

（2） "Auxiliary power unit" or "APU" means any device that provides electrical or mechanical energy, meeting the requirements of subdivision 1962. 1 （c）（2）, to a Type I. 5x or Type IIx vehicle, after the zero emission range has been fully depleted. A fuel fired heater does not qualify under this definition for an APU.

（3） "Battery electric vehicle" means any vehicle that operates solely by use of a battery or battery pack, or that is powered primarily through the use of an electric battery or battery pack but uses a flywheel or capacitor that stores energy produced by the electric motor or through regenerative braking to assist in vehicle operation.

（4） "Charge depletion range actual" or "R_{cda}" means the distance achieved by a hybrid electric vehicle on the urban driving cycle at the point when the zero-emission energy storage device is depleted of off-vehicle charge and regenerative braking derived energy.

（5） "Conventional rounding method" means to increase the last digit to be retained when the following digit is five or greater. Retain the last digit as is when the following digit is four or less.

（6） "East Region pool" means the combination Section 177 states east of the Mississippi River.

（7） "Electric drive system" means an electric motor and associated power electronics which provide acceleration torque to the drive wheels sometime during normal vehicle operation. This does not include components that could act as a motor, but are configured to act only as a generator or engine starter in a particular vehicle application.

(8) "Enhanced AT PZEV" means any model year 2009 through 2011 PZEV that has an allowance of 1. 0 or greater per vehicle without multipliers and makes use of a ZEV fuel. Enhanced AT PZEV means Transitional Zero Emission Vehicle.

(9) "Neighborhood electric vehicle" or "NEV" means a motor vehicle that meets the definition of Low-Speed Vehicle either in section 385. 5 of the Vehicle Code or in 49 CFR 571. 500 (as it existed on July 1, 2000), and is certified to zero-emission vehicle standards.

(10) "Placed in service" means having been sold or leased to an end-user and not to a dealer or other distribution chain entity, and having been individually registered for on-road use by the California DMV.

(11) "Proportional value" means the ratio of a manufacturer's California applicable sales volume to the manufacturer's Section 177 state applicable sales volume. In any given model year, the same applicable sale volume calculation method must be used to calculate proportional value.

(12) "Range Extended Battery Electric Vehicle" means a vehicle powered predominantly by a zero emission energy storage device, able to drive the vehicle for more than 75 all-electric miles, and also equipped with a backup APU, which does not operate until the energy storage device is fully depleted, and meeting requirements in subdivision 1962. 1 (d)(5)(G),

(13) "Regenerative braking" means the partial recovery of the energy normally dissipated into friction braking that is returned as electrical current to an energy storage device.

(14) "Section 177 state" means a state that is administering the California ZEV requirements pursuant to section 177 of the federal Clean Air Act (42 U. S. C. § 7507).

(15) "Transitional Zero Emission Vehicle" means a PZEV that has an allowance of 1. 0 or greater, and makes use of a ZEV fuel.

(16) "Type 0, I, I. 5, II, III, IV, and V ZEV" all have the meanings set forth in section 1962. 1 (d)(5)(A).

(17) "West Region pool" means the combination of Section 177 states west of the Mississippi River.

(18) "ZEV fuel" means a fuel that provides traction energy in on-road ZEVs. Examples of current technology ZEV fuels include electricity, hydrogen, and compressed air.

(j) *Abbreviations*. The following abbreviations are used in this section 1962. 1:

"AER" means all-electric range.

"APU" means auxiliary power unit.

"AT PZEV" means advanced technology partial zero-emission vehicle.

"CFR" means Code of Federal Regulations.

"DMV" means the California Department of Motor Vehicles.

"EAER" means equivalent all electric range.

"$EAER_{u40}$" means the urban equivalent all-electric range that a 40 mile R_{cda} plug-in hybrid electric vehicle achieves.

"FR" means Federal Register.

"HEV" means hybrid-electric vehicle.

"LDT" means light-duty truck.

"LDT1" means a light-truck with a loaded vehicle weight of 0 – 3750 pounds.

"LDT2" means a "LEV II" light-duty truck with a loaded vehicle weight of 3 751 pounds to a gross vehicle weight of 8 500 pounds, or a "LEV I" light-duty truck with a loaded vehicle weight of 3 751 – 5 750 pounds.

"LVM" means large volume manufacturer.

"MDV" means medium-duty vehicle.

"Non-Methane Organic Gases" or "NMOG" means the total mass of oxygenated and non-oxygenated hydrocarbon emissions.

"NEV" means neighborhood electric vehicle.

"NOx" means oxides of nitrogen.

"PC" means passenger car.

"PZEV" means partial allowance zero-emission vehicle, any vehicle that is delivered for sale in California and that qualifies for a partial ZEV allowance of at least 0. 2.

"R_{cda}" means urban charge depletion range actual.

"SAE" means Society of Automotive Engineers.

"SULEV" means super-ultra-low-emission-vehicle.

"TZEV" means transitional zero emission vehicle.

"Type I. 5x" means range extended 75 mile to 100 mile all electric range battery electric vehicle.

"Type IIx" means range extended 100 mile or greater all electric range battery electric vehicle.

"UDDS" means urban dynamometer driving cycle.

"UF" means utility factor.

"US06" means the US06 Supplemental Federal Test Procedure

"VMT" means vehicle miles traveled.

"ZEV" means zero-emission vehicle.

(k) *Severability*. Each provision of this section is severable, and in the event that any provision of this section is held to be invalid, the remainder of this article remains in full force and effect.

(l) *Public Disclosure*. Records in the Board's possession for the vehicles subject to the requirements of section 1962. 1 shall be subject to disclosure as public records as follows:

(1) Each manufacturer's annual production data and the corresponding credits per vehicle earned for ZEVs (including ZEV type), TZEVs, AT PZEVs, and PZEVs for the 2009 through 2017 model years; and

(2) Each manufacturer's annual credit balances for 2010 through 2017 years for:

(A) Each type of vehicle: ZEVs (minus NEVs), Type I. 5x, and Type IIx vehicles, NEVs, TZEVs, AT PZEVs, and PZEVs; and

(B) Advanced technology demonstration programs; and

(C) Transportation systems; and

(D) Credits earned under subdivision 1962. 1 (d)(5)(C), including credits acquired from, or transferred to another party.

Note: Authority cited: Sections 39600, 39601, 43013, 43018, 43101, 43104 and 43105, Health and Safety Code. Reference: Sections 38562, 39002, 39003, 39667, 43000, 43009. 5, 43013, 43018, 43018. 5, 43100, 43101, 43101. 5, 43102, 43104, 43105, 43106, 43204, 43205, 43205. 5 and 43206, Health and Safety Code.

§ 1962. 2 Zero-Emission Vehicle Standards for 2018 and Subsequent Model Year Passenger Cars, Light-Duty Trucks, and Medium-Duty Vehicles.

(a) *ZEV Emission Standard*. The Executive Officer shall certify new 2018 and subsequent model year passenger cars, light-duty trucks, and medium-duty vehicles as ZEVs, vehicles that produce zero exhaust emissions of any criteria pollutant (or precursor pollutant) or greenhouse gas, excluding emissions from air conditioning systems, under any possible operational modes or conditions.

(b) *Percentage ZEV Requirements*.

(1) *General ZEV Credit Percentage Requirement*.

(A) *Basic Requirement*. The minimum ZEV credit percentage requirement for each manufacturer is listed in the table below as the percentage of the PCs and LDTs, produced by the manufacturer and delivered for sale in California that must be ZEVs, subject to the conditions in this subdivision 1962. 2 (b). The ZEV requirement will be based on the annual NMOG production report for the appropriate model year.

Model Year	Credit Percentage Requirement
2018	4. 5%
2019	7. 0%
2020	9. 5%
2021	12. 0%
2022	14. 5%
2023	17. 0%
2024	19. 5%
2025 and subsequent	22. 0%

(B) *Calculating the Number of Vehicles to Which the Percentage ZEV Requirement is Applied*. For 2018 and subsequent model years, a manufacturer's production volume for the given model year will be based on the three-year average of the manufacturer's volume of PCs and LDTs, produced and delivered for sale in California in the prior second, third, and fourth model year [for example, 2019 model year ZEV requirements will be based on California production volume average of PCs and LDTs for the 2015 to 2017 model years]. This production averaging is used to determine ZEV requirements only, and has no effect on a manufacturer's size determination (eg. three-year average calculation method). In applying the ZEV requirement, a PC or LDT, that is produced by one manufacturer (e. g. , Manufacturer A), but is marketed in California by another manufacturer (e. g. , Manufacturer B) under the other manufacturer's (Manufacturer B) nameplate, shall be treated as having been produced by the marketing manufacturer (i. e. , Manufacturer B).

1. [*Reserved*]

2. [*Reserved*]

3. A manufacturer may apply to the Executive Officer to be permitted to base its ZEV obligation on the number of PCs and LDTs, produced by the manufacturer and delivered for sale in California that same model year (ie, same model-year calculation method) as an alternative to the three-year averaging of prior year production described above, for up to two model years, total, between model year 2018 and model year 2025. For the same model-year calculation method to be allowed, a manufacturer's application to the Executive Officer must show that their volume of PCs and LDTs produced and delivered for sale in California has decreased by at least 30 percent from the previous year due to circumstances that were unforeseeable and beyond their control.

(**C**) [*Reserved*]

(**D**) *Exclusion of ZEVs in Determining a Manufacturer's Sales Volume*. In calculating a manufacturer's applicable sales, using either method described in subdivision 1962. 2 (b)(1)(B), a manufacturer shall exclude the number of NEVs produced and delivered for sale in California by the manufacturer itself, or by a subsidiary in which the manufacturer has more than 33. 4% percent ownership interest.

(**2**) *Requirements for Large Volume Manufacturers*.

(**A**) [*Reserved*]

(**B**) [*Reserved*]

(**C**) [*Reserved*]

(**D**) [*Reserved*]

(**E**) *Requirements for Large Volume Manufacturers in 2018 and through 2025 Model Years*. LVMs must produce credits from ZEVs equal to minimum ZEV floor percentage requirement, as enumerated below. Manufacturers may fulfill the remaining ZEV requirement with credits from TZEVs, as enumerated below.

Model Years	Total ZEV Percent Requirement	Minimum ZEV floor	TZEVs
2018	4. 5%	2. 0%	2. 5%
2019	7. 0%	4. 0%	3. 0%
2020	9. 5%	6. 0%	3. 5%
2021	12. 0%	8. 0%	4. 0%
2022	14. 5%	10. 0%	4. 5%
2023	17. 0%	12. 0%	5. 0%
2024	19. 5%	14. 0%	5. 5%
2025	22. 0%	16. 0%	6. 0%

(**F**) *Requirements for Large Volume Manufacturers in Model Year 2026 and Subsequent*. In 2026 and subsequent model years, a manufacturer must meet a total ZEV credit percentage of 22%. The maximum portion of a manufacturer's credit percentage requirement that may be satisfied by TZEV credits is limited to 6% of the manufacturer's applicable California PC and LDT production volume. ZEV credits must satisfy the remainder of the manufacturer's requirement.

(**3**) *Requirements for Intermediate Volume Manufacturers*. For 2018 and subsequent model years, an intermediate volume manufacturer may meet all of its ZEV credit percentage requirement, under subdivision 1962. 2 (b), with credits from TZEV.

(**4**) *Requirements for Small Volume Manufacturers*. A small volume manufacturer is not required to meet the ZEV credit percentage requirements. However, a small volume manufacturer may earn, bank, market, and trade credits for the ZEVs and TZEVs it produces and delivers for sale in California.

(**5**) [*Reserved*]

(**6**) [*Reserved*]

(**7**) *Changes in Small Volume and Intermediate Volume Manufacturer Status*.

(**A**) *Increases in California Production Volume*. In 2018 and subsequent model years, if a small volume manufacturer's average California production volume exceeds 4 500 units of new PCs, LDTs, and MDVs based on the average number of vehicles produced and delivered for sale for the three previous consecutive model years (i. e. , total production volume exceeds 13 500 vehicles in a three-year period), for three consecutive averages, the manufacturer shall no longer be treated as a small volume manufacturer, and must comply with the ZEV requirements for intermediate volume manufacturers beginning with the next model year after the last model year of the third consecutive average. For example, if (a small volume) Manufacturer A exceeds 4 500 PCs, LDTs, and MDVs for their 2018 – 2020, 2019 – 2021, and 2020 – 2022 model year averages, Manufacturer A would be subject to intermediate volume requirements starting in 2023 model year.

If an intermediate volume manufacturer's average California production volume exceeds 20 000 units of new PCs, LDTs, and MDVs based on the average number of vehicles produced and delivered for sale for the three previous consecutive model years (i. e. , total production volume exceeds 60 000 vehicles in a three – year period), for three consecutive averages, the manufacturer shall no longer be treated as an intermediate volume manufacturer and shall comply with the ZEV requirements for large volume manufacturers beginning with the next model year after the last model year of the third consecutive average. For example, if (an intermediate volume) Manufacturer B exceeds 20 000 PCs, LDTs, and MDVs for its 2018 – 2020, 2019 – 2021, and 2020 – 2022 average, Manufacturer B would be subject to large volume manufacturer requirements starting in 2023 model year.

Any new requirement described in this subdivision will begin with the next model year after the last model year of the third consecutive averages when a manufacturer ceases to be a small or intermediate volume manufacturer in 2018 or subsequent years due to the aggregation requirements in majority ownership situations.

(B) *Decreases in California Production Volume*. If a manufacturer's average California production volume falls below 4 500 or 20 000 units of new PCs, LDT1 and 2s, and MDVs, based on the average number of vehicles produced and delivered for sale for the three previous consecutive model years, for three consecutive averages, the manufacturer shall be treated as a small volume or intermediate volume manufacturer, as applicable, and shall be subject to the requirements for a small volume or intermediate volume manufacturer beginning with the next model year. For example, if Manufacturer C falls below 20 000 PCs, LDTs, and MDVs for its 2019 – 2021, 2020 – 2022, and 2021 – 2023 averages, Manufacturer C would be subject to IVM requirements starting in 2024 model year.

(C) *Calculating California Production Volume in Change of Ownership Situations*. Where a manufacturer experiences a change in ownership in a particular model year, the change will affect application of the aggregation requirements on the manufacturer starting with the next model year. When a manufacturer is simultaneously producing two model years of vehicles at the time of a change of ownership, the basis of determining next model year must be the earlier model year. The manufacturer's small or intermediate volume manufacturer status for the next model year shall be based on the average California production volume in the three previous consecutive model years of those manufacturers whose production volumes must be aggregated for that next model year. For example, where a change of ownership during the 2019 calendar year occurs and the manufacturer is producing both 2019 and 2020 model year vehicles resulting in a requirement that the production volume of Manufacturer A be aggregated with the production volume of Manufacturer B, Manufacturer A's status for the 2020 model year will be based on the production volumes of Manufacturers A and B in the 2017 – 2019 model years. Where the production volume of Manufacturer A must be aggregated with the production volumes of Manufacturers B and C for the 2019 model year, and during that model year a change in ownership eliminates the requirement that Manufacturer B's production volume be aggregated with Manufacturer A's, Manufacturer A's status for the 2020 model year will be based on the production volumes of Manufacturers A and C in the 2017 – 2019 model years. In either case, the lead time provisions in subdivisions 1962. 2 (b)(7)(A) and (B) will apply.

(c) *Transitional Zero-Emission Vehicles* (*TZEV*).

(1) *Introduction*. This subdivision 1962. 2 (c) sets forth the criteria for identifying vehicles delivered for sale in California as TZEVs.

(2) *TZEV Requirements*. In order for a vehicle to be eligible to receive a ZEV allowance, the manufacturer must demonstrate compliance with all of the following requirements:

(A) *SULEV Standards*. Certify the vehicle to the 150 000 – mile SULEV 20 or 30 exhaust emission standards for PCs and LDTs in subdivision 1961. 2 (a)(1). Bi-fuel, fuel flexible and dual-fuel vehicles must certify to the applicable 150 000 – mile SULEV 20 or 30 exhaust emission standards when operating on both fuels. Manufacturers may certify 2018 and 2019 TZEVs to the 150 000 – mile SULEV exhaust emission standards for PCs and LDTs in subdivision 1961 (a)(1);

(B) *Evaporative Emissions*. Certify the vehicle to the evaporative emission standards in subdivision 1976 (b)(1)(G) or 1976 (b)(1)(E);

(C) *OBD*. Certify that the vehicle will meet the applicable on-board diagnostic requirements in sections 1968. 1 or 1968. 2, as applicable, for 150 000 miles; and

(D) *Extended Warranty*. Extend the performance and defects warranty period set forth in subdivisions 2037 (b)(2) and 2038 (b)(2) to 15 years or 150 000 miles, whichever occurs first except that the time period is to be 10 years for a zero-emission energy storage device used for traction power (such as a battery, ultracapacitor, or other electric storage device).

(3) *Allowances for TZEVs*

(A) *Zero-Emission Vehicle Miles Traveled TZEV Allowance Calculation.* A vehicle that meets the requirements of subdivision 1962. 2 (c) (2) and has zero-emission vehicle miles traveled (VMT), as defined by and calculated by the "California Exhaust Emission Standards and Test Procedures for 2018 and Subsequent Model Zero-Emission Vehicles and Hybrid Electric Vehicles, in the Passenger Car, Light-Duty Truck and Medium-Duty Vehicle Classes," adopted March 22, 2012, which is incorporated herein by reference, and measured as equivalent all electric range (EAER) capability will generate an allowance according to the following equation:

UDDS Test Cycle Range (AER)	Allowance
< 10 all electric miles	0. 00
≥ 10 all electric miles	TZEV Credit = [(0. 01) × EAER + 0. 30]
> 80 miles (credit cap)	1. 10

1. *Allowance for US06 Capability.* TZEVs with US06 all electric range capability (AER) of at least 10 miles shall earn an additional 0. 2 allowance. US06 test cycle range capability shall be determined in accordance with section G. 7. 5 of the "California Exhaust Emission Standards and Test Procedures for the 2018 and Subsequent Model Zero-Emission Vehicles, and Hybrid Electric Vehicles in the Passenger Car, Light-Duty Truck, and Medium Duty Vehicle Classes," adopted March 22, 2012, last amended December 6, 2012, which is incorporated herein by reference.

(B) [*Reserved*]

(C) [*Reserved*]

(D) [*Reserved*]

(E) *Credit for Hydrogen Internal Combustion Engine Vehicles.* A hydrogen internal combustion engine vehicle that meets the requirements of subdivision 1962. 2 (c) (2) and has a total range of at least 250 UDDS miles will earn an allowance of 0. 75, which may be in addition to allowances earned in subdivision 1962. 2 (c) (3) (A), and subject to an overall credit cap of 1. 25.

(d) *Qualification for Credits From ZEVs.*

(1) [*Reserved*]

(2) [*Reserved*]

(3) [*Reserved*]

(4) [*Reserved*]

(5) *Credits for 2018 and Subsequent Model Year ZEVs.*

(A) *ZEV Credit Calculations.* Credits from a ZEV delivered for sale are based on the ZEV's UDDS all electric range, determined in accordance with the "California Exhaust Emission Standards and Test Procedures for the 2018 and Subsequent Model Zero-Emission Vehicles, and Hybrid Electric Vehicles in the Passenger Car, Light-Duty Truck, and Medium Duty Vehicle Classes," adopted March 22, 2012, which is incorporated herein by reference, using the following equation:

$$ZEV\ Credit = (0.01) \times (UDDS\ range) + 0.50$$

1. A ZEV with less than 50 miles UDDS range will receive zero credits.

2. Credits earned under this provision 1962.2 (d)(5)(A) are be capped at 4 credits per ZEV.

(B) [*Reserved*]

(C) [*Reserved*]

(D) [*Reserved*]

(E) **1.** *Counting Specified ZEVs Placed in Service in a Section 177 State and in California.* Large volume manufacturers and intermediate volume manufacturers with credits earned from hydrogen fuel cell vehicles that are certified to the California ZEV standards applicable for the ZEV's model year, delivered for sale and placed in service in California or in a Section 177 state, may be counted towards compliance in California and in all Section 177 states with the percentage ZEV requirements in subdivision 1962.2 (b). The credits earned are multiplied by the ratio of a manufacturer's applicable production volume for a model year, as specified in subdivision 1962.2 (b)(1)(B), in the state receiving credit to the manufacturer's applicable production volume as specified in subdivision 1962.2 (b)(1)(B), for the same model year in California (hereafter, "proportional value"). Credits generated from ZEV placement in a Section 177 state will be earned at the proportional value in the Section 177 state, and earned in California at the full value specified in subdivision 1962.2 (d)(5)(A).

1. *Optional Section 177 State Compliance Path.*

a. *Reduced ZEV and TZEV Percentages*. Large volume manufacturers and intermediate volume manufacturers that have fully complied with the optional Section 177 state compliance path requirements in subdivision 1962. 1 (d)(5)(E) 3. are allowed to meet ZEV percentage requirements and optional TZEV percentages reduced from the minimum ZEV floor percentages and TZEV percentages in subdivision 1962. 2 (b)(2)(E) in each Section 177 state equal to the following percentages of their sales volume determined under subdivision 1962. 2 (b)(1)(B):

ZEVs

Model Year	*2018*	*2019*	*2020*	*2021*
Existing Minimum ZEV Floor	2. 00%	4. 00%	6. 00%	8. 00%
Section 177 State Adjustment for Optional Compliance Path	62. 5%	75%	87. 5%	100%
Minimum Section 177 State ZEV Requirement	1. 25%	3. 00%	5. 25%	8. 00%

TZEVs

Model Year	*2018*	*2019*	*2020*	*2021*
Existing TZEV Percentage	2. 50%	3. 00%	3. 50%	4. 00%
Section 177 State Adjustment for Optional Compliance Path	90. 00%	100%	100%	100%
New Section 177 State TZEV Percentage	2. 25%	3. 00%	3. 50%	4. 00%

Total Percent Requirement

Model Year	*2018*	*2019*	*2020*	*2021*
New Total Section 177 State Optional Requirements	3. 50%	6. 00%	8. 75%	12. 00%

i. *Tading and Transferring ZEV and TZEV Credits Within West Region Pool and East Region Pool*. Manufacturers that have fully complied with the optional section 177 state compliance path requirements in subdivision 1962. 1(d)(5)(E)3. may trade or transfer specified model year ZEV and TZEV credits within the East Region pool to meet the same model year requirements in subdivision 1962. 2(d)(5)(E)2. a, and will incur no premium on their credit values. For example, for a manufacturer to make up a 2019 model year shortfall of 100 credits in State W, the manufacturer may transfer 100 (2019 model year) ZEV credits from State Z, within the East Region pool.

ii. *Trading and Transferring ZEV and TZEV Credits between the West Region Pool and East Region Pool*. Manufacturers that have fully complied with the optional section 177 state compliance path requirements in subdivision 1962. 1 (d)(5)(E) 3. may trade or transfer specified model year ZEV and TZEV credits to meet the same model year requirements in subdivision 1962. 2 (d)(5)(E) 2. a. between the West Region pool and the East Region pool; however, any credits traded will incur a premium of 30% of their value. For example, in order for a manufacturer to make up a 2019 model year shortfall of 100 credits in the West Region Pool, the manufacturer may transfer 130 (2019 model year) credits from the East Region Pool. No credits may be traded or transferred to the East Region pool or West Region pool from a manufacturer's California ZEV bank, or from the East Region pool or West Region pool to a manufacturer's California ZEV bank.

b. *Reporting Requirements*. On an annual basis, by May 1st of the calendar year following the close of a model year, each manufacturer that elects the optional section 177 state compliance path under subdivision 1962. 1 (d)(5)(E) 3, shall submit, in writing, to the Executive Officer and each section 177 state a report, including an itemized list, that indicates where vehicles have been placed within the East Region pool and within the West Region pool. The itemized list shall include the following:

i. The manufacturer's total applicable volume of PCs and LDTs delivered for sale in each section 177 state within the regional pool, as determined under subdivision 1962. 2 (b)(1)(B).

ii. Make, model, vehicle identification number credit earned, and section 177 state where delivery for sale of each TZEV and ZEV occurred to meet manufacturer's requirements under subdivision 1962. 2 (d)(5)(E) 2. a.

c. *Failure to Meet Optional Section* 177 *State Compliance Path Requirements*. A manufacturer that elects the optional section 177 state compliance path under subdivision 1962. 1(d)(5)(E)3, and does not meet the modified percentages in subdivision 1962. 2 (d)(5)(E)2. a. in a model year or make up their deficit within the specified time and with the specified credits allowed by subdivision 1962. 2(g)(7)(A) in all section 177 states of the applicable pool, shall be treated as subject to the total ZEV percentage requirements in section 1962. 2(b) for the 2018 through 2021 model years in each section 177 state, and the pooling provisions in subdivision 1962. 2(d)(5)(E)2. a. shall not apply. Any transfers of ZEV or TZEV credits between section 177 states will be null and void if a manufacturer fails to comply, and ZEV or TZEV credits will return to the section 177 state in which the credits were earned. Penalties shall be calculated separately by each section 177 state where a manufacturer fails to make up the ZEV deficits by the end of the 2018 model year.

d. The provisions of section 1962. 2 shall apply to a manufacturer electing the optional section 177 state compliance path, except as specifically modified by this subdivision 1962. 2 (d)(5)(E) 2.

(F) NEVs. NEVs must meet the following to be eligible for 0. 15 credits:

1. Specifications. A NEV earns credit when it meets all the following specifications:

a. Acceleration. The vehicle has a 0 – 20 mph acceleration of 6. 0 seconds or less when operating with a payload of at least 332 pounds and starting with the battery at a 50% state of charge.

b. Top Speed. The vehicle has a minimum top speed of 20 mph when operating with a payload of at least 332 pounds and starting with the battery at a 50% state of charge. The vehicle's top speed shall not exceed 25 mph when tested in accordance with 49 CFR 571. 500 (68 FR 43972, July 25, 2003).

c. Constant Speed Range. The vehicle has a minimum 25 – mile range when operating at constant top speed with a payload of least 332 pounds and starting with the battery at 100% state of charge.

2. Battery Requirement. A NEV must be equipped with one or more sealed, maintenance-free batteries.

3. Warranty Requirement. A NEV drive train, including battery packs, must be covered for a period of at least 24 months. The first 6 months of the NEV warranty period must be covered by a full warranty; the remaining warranty period may be optional extended warranties (available for purchase) and may be prorated. If the extended warranty is prorated, the percentage of the battery pack's original value to be covered or refunded must be at least as high as the percentage of the prorated coverage period still remaining. For the purpose of this computation, the age of the battery pack must be expressed in intervals no larger than three months. Alternatively, a manufacturer may cover 50 percent of the original value of the battery pack for the full period of the extended warranty.

Prior to credit approval, the Executive Officer may request that the manufacturer provide copies of representative vehicle and battery warranties.

4. *NEV Charging Requirements*. A NEV must meet charging requirements specific in subdivision 1962. 3 (c) (2) .

(G ） *BEVx*. A BEVx must meet the following in order to receive credit, based on its all electric UDDS Range, through subdivision 1962. 2 (d) (5) (A) :

1. *Emissions Requirements*. BEVxs must meet all TZEV requirements, specified in subdivision 1962. 2 (c) (2) (A) through (D) .

2. *APU Operation*. The vehicle's UDDS range after the APU first starts and enters "charge sustaining hybrid operation" must be less than or equal to the vehicle's UDDS all-electric test range prior to APU start. The vehicle's APU cannot start under any user-selectable driving mode unless the energy storage system used for traction power is fully depleted.

3. *Minimum Zero Emission Range Requirements*. BEVxs must have a minimum of 75 miles UDDS all electric range.

(e) 〔*Reserved*〕

(f) 〔*Reserved*〕

(g) *Generation and Use of Credits*; *Calculation of Penalties*

(1) *Introduction*. A manufacturer that produces and delivers for sale in California ZEVs or TZEVs in a given model year exceeding the manufacturer's ZEV requirement set forth in subdivision 1962. 2 (b) shall earn ZEV credits in accordance with this subdivision 1962. 2 (g) .

(2) *ZEV Credit Calculations*.

(A) *Credits from ZEVs*. The amount of credits earned by a manufacturer in a given model year from ZEVs shall be expressed in units of credits, and shall be equal to the number of credits from ZEVs produced and delivered for sale in California that the manufacturer applies towards meeting the ZEV requirements, or, if applicable, requirements specified under subdivision 1962. 2 (d) (5) (E) 2. a. for the model year subtracted from the number of ZEVs produced and delivered for sale in California by the manufacturer in the model year.

(B) *Credits from TZEVs*. The amount of credits earned by a manufacturer in a given model year from TZEVs shall be expressed in units of credits, and shall be equal to the total number of TZEVs produced and delivered for sale in California that the manufacturer applies towards meeting its ZEV requirement, or, if applicable, requirements specified under subdivision 1962. 2 (d) (5) (E) 2. a. for the model year subtracted from the total number of ZEV allowances from TZEVs produced and delivered for sale in California by the manufacturer in the model year.

(C) *Separate Credit Accounts*. Credits from a manufacturer's ZEVs, BEVxs, TZEVs, and NEVs shall each be maintained in separate accounts.

(D) *Rounding Credits*. ZEV credits and debits shall be rounded to the nearest 1/100th only on the final credit and debit totals using the conventional rounding method.

(3) *ZEV Credits for MDVs*. Credits from ZEVs and TZEVs classified as MDVs, may be counted toward the ZEV requirement for PCs and LDTs, and included in the calculation of ZEV credits as specified in this subdivision 1962.2(g) if the manufacturer so specifies.

(4) *ZEV Credits for Advanced Technology Demonstration Programs*.

(A) [*Reserved*]

(B) *ZEVs*. ZEVs, including BEVxs, excluding NEVs, placed in a small or intermediate volume manufacturer's California advanced technology demonstration program for a period of two or more years, may earn ZEV credits even if the vehicle is not "delivered for sale" or registered with the California DMV. To earn such credits, the manufacturer must demonstrate to the reasonable satisfaction of the Executive Officer that the vehicles will be regularly used in applications appropriate to evaluate issues related to safety, infrastructure, fuel specifications or public education, and that for 50 percent or more of the first two years of placement the vehicle will be operated in California. Such a vehicle is eligible to receive the same credit that it would have earned if delivered for sale, and for fuel cell vehicles, placed in service. To determine vehicle credit, the model year designation for a demonstration vehicle shall be consistent with the model year designation for conventional vehicles placed in the same timeframe. Manufacturers may earn credit for up to 25 vehicles per model, per Section 177 state, per year under this subdivision 1962.2 (g) (4). A manufacturer's vehicles in excess of the 25 - vehicle cap will not be eligible for advanced technology demonstration program credits.

(5) *ZEV Credits for Transportation Systems*.

(A) [*Reserved*]

(B) [*Reserved*]

(C) *Cap on Use of Transportation System Credits*.

1. *ZEVs*. Transportation system credits earned or allocated by ZEVs or BEVxs pursuant to subdivision 1962.1(g)(5), not including any credits earned by the vehicle itself, may be used to satisfy up to one-tenth of a manufacturer's ZEV obligation in any given

model year, and may be used to satisfy up to one-tenth of a manufacturer's ZEV obligation which must be met with ZEVs, as specified in subdivision 1962.2(b)(2)(E) or, if applicable, requirements specified under subdivision 1962.2(d)(5)(E) 2. a.

 2. *TZEVs.* Transportation system credits earned or allocated by TZEVs pursuant to subdivision 1962.1 (g)(5), not including all credits earned by the vehicle itself, may be used to satisfy up to one-tenth of the portion of a manufacturer's ZEV obligation that may be met with TZEVs, or, if applicable, the portion of a manufacturer's obligation that may be met with TZEVs specified under subdivision 1962.2 (d)(5)(E) 2. a. in any given model year, but may only be used in the same manner as other credits earned by vehicles of that category.

 (6) *Use of ZEV Credits.* A manufacturer may meet the ZEV requirements in a given model year by submitting to the Executive Officer a commensurate amount of ZEV credits, consistent with subdivision 1962.2 (b). Credits in each of the categories may be used to meet the requirement for that category as well as the requirements for lesser credit earning ZEV categories, but shall not be used to meet the requirement for a greater credit earning ZEV category, except for discounted PZEV and AT PZEV credits. For example, credits produced from TZEVs may be used to comply with the portion of the requirement that may be met with credits from TZEV, but not with the portion that must be satisfied with credits from ZEVs. These credits may be earned previously by the manufacturer or acquired from another party.

 (A) *Use of Discounted PZEV and AT PZEV Credits and NEV Credits.* For model years 2018 through 2025, discounted PZEV and AT PZEV credits, and NEV credits may be used to satisfy up to one-quarter of the portion of a manufacturer's requirement that can be met with credits from TZEVs, or, if applicable, the portion of a manufacturer's obligation that may be met with TZEVs specified under subdivision 1962.2 (d)(5)(E) 2. a. . Intermediate volume manufacturers may fulfill their entire requirement with discounted PZEV and AT PZEV credits, and NEV credits in model years 2018 and 2019. These credits may be earned previously by the manufacturer or acquired from another party. Discounted PZEV and AT PZEV credits may no longer be used after model year 2025 compliance.

 (B) *Use of BEVx Credits.* BEVx credits may be used to satisfy up to 50% of the portion of a manufacturer's requirement that must be met with ZEV credits.

 (C) *GHG-ZEV Over Compliance Credits.*

 1. *Application.* Manufacturers may apply to the Executive Officer, no later than

December 31, 2016, to be eligible for this subdivision 1962.2 (g)(6)(C), based on the following qualifications:

    a. A manufacturer must have no model year 2017 compliance debits and no outstanding debits from all previous model year compliance with sections 1961.1 and 1961.3, or must have demonstrated compliance with the National greenhouse gas program as allowed by subdivisions 1961.1 (a)(1)(A)(ii) and 1961.3 (c); and

    b. A manufacturer must have no model year 2017 compliance debits and no outstanding debits from all previous model year compliance with section 1962.1; and

    c. A manufacturer must submit documentation of its projected product plans to show over compliance with the manufacturer's section 1961.3 requirements, or over compliance with National greenhouse gas program requirements as allowed by subdivision 1961.3 (c), by at least 2.0 gCO_2/mile in each model year through the entire 2018 through 2021 model year period, and its commitment to do so in each year.

    2. *Credit Generation and Calculation*. Manufacturers must calculate their over compliance with section 1961.3 requirements, or over compliance with the National greenhouse gas program requirements as allowed by subdivision 1961.3 (c), for model years 2018 through 2021 based on compliance with the previous model year standard. For example, to generate credits for this subdivision 1962.2 (g)(6)(C) for model year 2018, manufacturers would calculate credits based on model year 2017 compliance with section 1961.3, or over compliance with the National greenhouse gas program as allowed by subdivision 1961.3 (c).

    a. At least 2.0 gCO_2/mile over compliance with section 1961.3, or over compliance with the National greenhouse gas program as allowed by subdivision 1961.3 (c), is required in each year and the following equation must be used to calculate the amount of ZEV credits earned for purposes of this subdivision 1962.2 (g)(6)(C), and:

[(*Manufacturer US PC and LDT Sales*) x (gCO_2/*mile below manufacturer GHG standard for a given model year*)] / (*Manufacturer GHG standard for a given model year*)

    b. Credits earned under subdivision 1961.3 (a)(9), or credits earned under 40 CFR, part 86, Subpart S §86.1866 – 12 (a), §86.1866 – 12 (b), or §86.1870 – 12, may not be included in the calculation of gCO_2/mile credits for use in the above equation in subdivision a. All ZEVs included in the calculation above must include upstream emission values found in section 1961.3.

    c. Banked gCO_2/mile credits earned under sections 1961.1 and 1961.3, or under the National greenhouse gas program requirements as allowed by subdivision 1961. 3 (c) , from previous model years or from other manufacturers may not be included in the calculation of gCO_2/mile credits for use in the above equation in subdivision a.

3. *Use of GHG-ZEV Over Compliance Credits*. A manufacturer may use no more than the percentage enumerated in the table below to meet either the total ZEV requirement nor the portion of their ZEV requirement that must be met with ZEV credits, with credits earned under this subdivision 1962. 2 (g) (6) (C) .

2018	*2019*	*2020*	*2021*
50%	50%	40%	30%

Credits earned in any given model year under this subdivision 1962. 2 (g) (6) (C) may only be used in the applicable model year and may not be used in any other model year.

gCO_2/mile credits used to calculate GHG-ZEV over compliance credits under this provision must also be removed from the manufacturer's GHG compliance bank, and cannot be banked for future compliance toward section 1961. 3, or towards compliance with the National greenhouse gas program requirements as allowed by subdivision 1961. 3(c) .

4. *Reporting Requirements*. Annually, manufacturers are required to submit calculations of credits for this subdivision 1962. 2(g) (6) (C) for the model year, any remaining credits/debits from previous model years under section 1961. 3 or under the National greenhouse gas program requirements as allowed by subdivision 1961. 3(c) , and projected credits/debits for future years through 2021 under section 1961. 3 or under the National greenhouse gas program requirements as allowed by subdivision 1961. 3(c) and this subdivision 1962. 2(g) (6) (C) .

If a manufacturer, who has been granted the ability to generate credits under this subdivision 1962. 2 (g) (6) (C) , fails to over comply by at least 2. 0 gCO_2/mile in any one year, the manufacturer will be subject to the full ZEV requirements for the model year and future model years, and will not be able to earn credits for any other model year under this subdivision 1962. 2 (g) (6) (C) .

(7) *Requirement to Make Up a ZEV Deficit*.

(A) *General*. A manufacturer that produces and delivers for sale in California fewer ZEVs than required in a given model year shall make up the deficit by the next model

year by submitting to the Executive Officer a commensurate amount of ZEV credits. The a-
mount of ZEV credits required to be submitted shall be calculated by [i] adding the number
of credits from ZEVs produced and delivered for sale in California by the manufacturer for
the model year to the number of credits from TZEVs produced and delivered for sale in California by the manufacturer for the model year (for a LVM, not to exceed that permitted under subdivision 1962. 2 (b)(2)), and [ii] subtracting that total from the number of credits required to be produced and delivered for sale in California by the manufacturer for the
model year. BEVx, TZEV, NEV, or converted AT PZEV and PZEV credits are not allowed
to be used to fulfill a manufacturer's ZEV deficit; only credits from ZEVs may be used to
fulfill a large volume manufacturer's ZEV deficit.

(8) *Penalty for Failure to Meet ZEV Requirements*. Any manufacturer that fails to
produce and deliver for sale in California the required number of ZEVs and submit an appropriate amount of credits and does not make up ZEV deficits within the specified time allowed
by subdivision 1962. 2(g)(7)(A) shall be subject to the Health and Safety Code section
43211 civil penalty applicable to a manufacturer that sells a new motor vehicle that does not
meet the applicable emission standards adopted by the state board. The cause of action shall
be deemed to accrue when the ZEV deficit is not balanced by the end of the specified time
allowed by subdivision 1962. 2(g)(7)(A). For the purposes of Health and Safety Code
section 43 211, the number of vehicles not meeting the state board's standards shall be e-
qual to the manufacturer's credit deficit, rounded to the nearest $1/100^{th}$, calculated according to the following equation, provided that the percentage of a manufacturer's ZEV requirement for a given model year that may be satisfied with TZEVs or credit from such vehicles
may not exceed the percentages permitted under subdivision 1962. 2(b)(2):

(No. of ZEV credits required to be generated for the model year) − (Amount of
credits submitted for compliance for the model year)

(h) *Test Procedures*.

(1) *Determining Compliance*. The certification requirements and test procedures for
determining compliance with this section 1962. 2 are set forth in "California Exhaust Emission Standards and Test Procedures for 2018 and Subsequent Model Zero-Emission Vehicles
and Hybrid Electric Vehicles, in the Passenger Car, Light-Duty Truck and Medium-Duty
Vehicle Classes," adopted March 22, 2012, last amended December 6, 2012, which is incorporated herein by reference.

(2) *NEV Compliance*. The test procedures for determining compliance with subdivi-

sion 1962. 1(d)(5)(F)1. are set forth in ETA-NTP002 (revision 3) "Implementation of SAE Standard J1666 May 93: Electric Vehicle Acceleration, Gradeability, and Deceleration Test Procedure" (December 1, 2004), and ETA-NTP004 (revision 3) "Electric Vehicle Constant Speed Range Tests" (February 1, 2008), both of which are incorporated by reference herein.

(i) *ZEV-Specific Definitions*. The following definitions apply to this section 1962. 2.

(1) "Auxiliary power unit" or "APU" means any device that provides electrical or mechanical energy, meeting the requirements of subdivision 1962. 2(c)(2), to a BEVx, after the zero emission range has been fully depleted. A fuel fired heater does not qualify under this definition for an APU.

(2) "Charge depletion range actual" or "R_{cda}" means the distance achieved by a hybrid electric vehicle on the urban driving cycle at the point when the zero-emission energy storage device is depleted of off-vehicle charge and regenerative braking derived energy.

(3) "Conventional rounding method" means to increase the last digit to be retained when the following digit is five or greater. Retain the last digit as is when the following digit is four or less.

(4) "Discounted PZEV and AT PZEV credits" means credits earned under section 1962 and 1962. 1 by delivery for sale of PZEVs and AT PZEVs, discounted according to subdivision 1962. 1(g)(2)(F).

(5) "East Region pool" means the combination of Section 177 states east of the Mississippi River.

(6) "Energy storage device" means a storage device able to provide the minimum power and energy storage capability to enable engine stop/start capability, traction boost, regenerative braking, and (nominal) charge sustaining mode driving capability. In the case of TZEVs, a minimum range threshold relative to certified, new-vehicle range capability is not specified or required.

(7) "Hydrogen fuel cell vehicle" means a ZEV that is fueled primarily by hydrogen, but may also have off-vehicle charge capability.

(8) "Hydrogen internal combustion engine vehicle" means a TZEV that is fueled exclusively by hydrogen.

(9) "Majority ownership situations" means when one manufacturer owns another manufacturer more than 33. 4%, for determination of size under CCR Section 1900.

(10) "Manufacturer US PC and LDT Sales" means a manufacturer's total passenger

car and light duty truck (up to 8 500 pounds loaded vehicle weight) sales sold in the United States of America in a given model year.

(11) "Neighborhood electric vehicle" or "NEV" means a motor vehicle that meets the definition of Low-Speed Vehicle either in section 385. 5 of the Vehicle Code or in 49 CFR 571. 500 (as it existed on July 1, 2000), and is certified to zero-emission vehicle standards.

(12) "Placed in service" means having been sold or leased to an end-user and not to a dealer or other distribution chain entity, and having been individually registered for on-road use by the California DMV.

(13) "Proportional value" means the ratio of a manufacturer's California applicable sales volume to the manufacturer's Section 177 state applicable sales volume. In any given model year, the same applicable sales volume calculation method must be used to calculate proportional value.

(14) "Range Extended Battery Electric Vehicle" or "BEVx" means a vehicle powered predominantly by a zero emission energy storage device, able to drive the vehicle for more than 75 all-electric miles, and also equipped with a backup APU, which does not operate until the energy storage device is fully depleted, and meeting requirements in subdivision 1962. 2 (d)(5)(G).

(15) "Section 177 state" means a state that is administering the California ZEV requirements pursuant to section 177 of the federal Clean Air Act (42 U. S. C. § 7507).

(16) "Transitional zero emission vehicle" or "TZEV" means a vehicle that meets all the criteria of subdivision 1962. 2 (c)(2) and qualifies for an allowance in subdivision 1962. 2 (c)(3)(D) or (E).

(17) "West Region pool" means the combination of section 177 states west of the Mississippi River.

(18) "Zero emission vehicle" or "ZEV" means a vehicle that produces zero exhaust emissions of any criteria pollutant (or precursor pollutant) or greenhouse gas under any possible operational modes or conditions.

(19) "Zero emission vehicle fuel" means a fuel that provides traction energy in on-road ZEVs. Examples of current technology ZEV fuels include electricity, hydrogen, and compressed air.

(j) *Abbreviations*. The following abbreviations are used in this section 1962. 2:

"AER" means all-electric range.

"APU" means auxiliary power unit.

"AT PZEV" means advanced technology partial zero-emission vehicle.

"BEVx" means range extended battery electric vehicle.

"CFR" means Code of Federal Regulations.

"CO_2" means carbon dioxide.

"DMV" means the California Department of Motor Vehicles.

"EAER" means equivalent all-electric range.

"FR" means Federal Register.

"g" means grams.

"HEV" means hybrid-electric vehicle.

"LDT" means light-duty truck.

"LDT1" means a light-truck with a loaded vehicle weight of 0 – 3750 pounds.

"LDT2" means a "LEV II" light-duty truck with a loaded vehicle weight of 3 751 pounds to a gross vehicle weight of 8 500 pounds, or a "LEV I" light-duty truck with a loaded vehicle weight of 3 751 – 5 750 pounds.

"LVM" means large volume manufacturer.

"MDV" means medium-duty vehicle.

"NMOG" means non-methane organic gases, or the total mass of oxygenated and non-oxygenated hydrocarbon emissions.

"NEV" means neighborhood electric vehicle.

"NOx" means oxides of nitrogen.

"PC" means passenger car.

"PZEV" means partial allowance zero-emission vehicle.

"SAE" means Society of Automotive Engineers.

"SULEV" means super-ultra-low-emission-vehicle.

"TZEV" means transitional zero emission vehicle.

"UDDS" means urban dynamometer driving cycle.

"US" means United States of America.

"US06" means the US06 Supplemental Federal Test Procedure.

"VMT" means vehicle miles traveled.

"ZEV" means zero-emission vehicle.

(k) *Severability*. Each provision of this section is severable, and in the event that any provision of this section is held to be invalid, the remainder of this article remains in full force and effect.

(1) ***Public Disclosure***. Records in the Board's possession for the vehicles subject to the requirements of section 1962. 2 shall be subject to disclosure as public records as follows:

(1) Each manufacturer's annual production data and the corresponding credits per vehicle earned for ZEVs and TZEVs for the 2018 and subsequent model years; and

(2) Each manufacturer's annual credit balances for 2018 and subsequent years for:

(A) Each type of vehicle: ZEV (minus NEV), BEVx, NEV, TZEV, and discounted PZEV and AT PZEV credits; and

(B) Advanced technology demonstration programs; and

(C) Transportation systems; and

(D) Credits earned under section 1962. 2 (d) (5) (A), including credits acquired from, or transferred to another party, and the parties themselves.

Note: Authority cited: Sections 39600, 39601, 43013, 43018, 43101, 43104 and 43105, Health and Safety Code. Reference: Sections 38562, 39002, 39003, 39667, 43000, 43009. 5, 43013, 43018, 43018. 5, 43100, 43101, 43101. 5, 43102, 43104, 43105, 43106, 43107, 43204 and 43205. 5, Health and Safety Code.

§ 1962. 3 Electric Vehicle Charging Requirements.

(a) *Applicability.* This section applies to:

(1) all battery electric vehicles, range extended battery electric vehicles, except for model year 2006 through 2013 neighborhood electric vehicles, that qualify for ZEV credit under section 1962. 1 and 1962. 2; and

(2) all hybrid electric vehicles that are capable of being recharged by a battery charger that transfers energy from the electricity grid to the vehicle for purposes of recharging the vehicle traction battery.

(b) *Definitions.*

(1) The definitions in section 1962. 1 and 1962. 2 apply to this section.

(c) *Requirements.*

(1) Beginning with the 2006 model year, all vehicles identified in subdivision (a) must be equipped with a conductive charger inlet and charging system which meets all the specifications applicable to AC Level 1 and Level 2 charging contained in Society of Automotive Engineers (SAE) Surface Vehicle Recommended Practice SAE J1772 REV JAN 2010, SAE Electric Vehicle and Plug in Hybrid Electric Vehicle Conductive Charger Coupler, which is incorporated herein by reference. All such vehicles must also be equipped with an on-board charger with a minimum output of 3. 3 kilowatts, or, sufficient power to enable a complete charge in less than 4 hours.

(2) A manufacturer may apply to the Executive Officer for approval to use an alternative to the AC inlet described in subdivision (c) (1), provided that the following conditions are met:

(A) each vehicle is supplied with a rigid adaptor that would enable the vehicle to meet all of the remaining system and on-board charger requirements described in subdivision (c) (1); and

(B) the rigid adaptor and alternative inlet must be tested and approved by a Nationally Recognized Testing Laboratory (NRTL).

Note: Authority cited: Sections 39600, 39601, 43013, 43018, 43101, 43104 and 43105, Health and Safety Code. Reference: Sections 38562, 39002, 39003, 39667, 43000, 43009. 5, 43013, 43018, 43018. 5, 43100, 43101, 43101. 5, 43102, 43104, 43105, 43106, 43107, 43204 and 43205. 5, Health and Safety Code.

The California Low-Emission Vehicle III Regulations

§ 1961. 2 Exhaust Emission Standards and Test Procedures – 2015 and Subsequent Model Passenger Cars, Light-Duty Trucks, and Medium-Duty Vehicles.

Introduction. This section 1961. 2 contains the California "LEV III" exhaust emission standards for 2015 and subsequent model year passenger cars, light-duty trucks, and medium-duty vehicles. A manufacturer must demonstrate compliance with the exhaust standards in subsection (a) applicable to specific test groups, and with the composite phase-in requirements in subsection (b) applicable to the manufacturer's entire fleet.

Before the 2015 model year, a manufacturer that produces vehicles that meet the standards in subsection (a) has the option of certifying the vehicles to those standards, in which case the vehicles will be treated as LEV III vehicles for purposes of the fleet-wide phase-in requirements. Similarly, 2015 – 2019 model-year vehicles may be certified to the "LEV II" exhaust emission standards in subsection 1961 (a) (1), in which case the vehicles will be treated as LEV II vehicles for purposes of the fleet-wide phase-in requirements.

A manufacturer has the option of certifying engines used in incomplete and diesel medium-duty vehicles with a gross vehicle weight rating of greater than 10 000 lbs. GVW to the heavy-duty engine standards and test procedures set forth in title 13, CCR, subsections 1956. 8 (c) and (h). All medium-duty vehicles with a gross vehicle weight rating of less than or equal to 10 000 lbs. GVW, including incomplete otto-cycle medium-duty vehicles and medium-duty vehicles that use diesel cycle engines, must be certified to the LEV III chassis standards and test procedures set forth in this section 1961. 2 in 2020 and subsequent model years.

Pooling Provision.

For each model year, a manufacturer must demonstrate compliance with this section 1961.2 based on one of two options applicable throughout the model year, either:

Option 1: the total number of passenger cars, light-duty trucks, and medium-duty vehicles that are certified to the California exhaust emission standards in subsection (a) and subsection 1961(a)(1), and are produced and delivered for sale in California; or

Option 2: the total number of passenger cars, light-duty trucks, and medium-duty vehicles that are certified to the California exhaust emission standards in subsection (a) and subsection 1961(a)(1), and are produced and delivered for sale in California, the District of Columbia, and all states that have adopted California's criteria pollutant emission standards set forth in this section 1961.2 for that model year pursuant to section 177 of the federal Clean Air Act (42 U.S.C. § 7507).

A manufacturer that selects compliance Option 2 must notify the Executive Officer of that selection in writing prior to the start of the applicable model year or must comply with Option 1. Once a manufacturer has selected compliance Option 2, that selection applies unless the manufacturer selects Option 1 and notifies the Executive Officer of that selection in writing before the start of the applicable model year.

When a manufacturer is demonstrating compliance using Option 2 for a given model year, the term "in California" as used in this section 1961.2 means California, the District of Columbia, and all states that have adopted California's criteria pollutant emission standards set forth in this section 1961.2 for that model year pursuant to section 177 of the federal Clean Air Act (42 U.S.C. § 7507).

(a) ***Exhaust Emission Standards.***

(1) *"LEV III" Exhaust Standards.* The following standards are the maximum exhaust emissions for the full useful life from new 2015 and subsequent model year "LEV III" passenger cars, light-duty trucks, and medium-duty vehicles, including fuel-flexible, bi-fuel and dual-fuel vehicles when operating on the gaseous or alcohol fuel they are designed to use. 2015 – 2019 model-year LEV II LEV vehicles may be certified to the 150 000 mile NMOG + NOx emission standards for LEV160, LEV395, or LEV630, as applicable, in this subsection (a)(1) and the corresponding NMOG + NOx numerical values in subsection (a)(4), in lieu of the separate NMOG and NOx exhaust emission standards in subsection 1961(a)(1) and the corresponding NMOG numerical values in subsection 1961(a)(4) and LEV II ULEV vehicles may be certified to the 150 000 mile NMOG + NOx emission

standards for ULEV125, ULEV340, or ULEV570, as applicable, in this subsection (a) (1) and the corresponding NMOG + NOx numerical values in subsection (a) (4), in lieu of the separate NMOG and NOx exhaust emission standards in subsection 1961 (a) (1) and the corresponding NMOG numerical values in subsection 1961 (a) (4). 2015 – 2019 model-year LEV II SULEV vehicles that receive a partial ZEV allowance in accordance with the "California Exhaust Emission Standards and Test Procedures for 2009 through 2017 Model Zero-Emission Vehicles and Hybrid Electric Vehicles, in the Passenger Car, Light-Duty Truck and Medium-Duty Vehicle Classes" and 2015 – 2016 model year vehicles that are allowed to certify to LEV II SULEV standards using "carryover" of emission test data under the provisions in subsection (b) (2) may be certified to the 150 000 mile NMOG + NOx emission standards for SULEV30, SULEV170, or SULEV230, as applicable, in this subsection (a) (1) and the corresponding NMOG + NOx numerical values in subsection (a) (4), in lieu of the separate NMOG and NOx exhaust emission standards in subsection 1961 (a) (1) and the corresponding NMOG numerical values in subsection 1961 (a) (4). LEV II SULEV vehicles that do not either (1) receive a partial ZEV allowance or (2) certify to LEV II SULEV standards in the 2015 – 2016 model years using "carryover" of emission test data may not certify to combined NMOG + NOx standards. LEV II vehicles that certify to combined NMOG + NOx standards will be treated as LEV II vehicles for purposes of the fleet-wide phase-in requirements.

LEV III Exhaust Mass Emission Standards for New 2015 and Subsequent Model Passenger Cars, Light-Duty Trucks, and Medium-Duty Vehicles						
Vehicle Type	*Durability Vehicle Basis (mi)*	*Vehicle Emission Category*[2]	*NMOG + Oxides of Nitrogen*[4] *(g/mi)*	*Carbon Monoxide (g/mi)*	*Formal-dehyde (mg/mi)*	*Particulates*[1] *(g/mi)*
All PCs; LDTs 8 500 lbs. GVWR or less; MDPVs Vehicles in this category are tested at their loaded vehicle weight	150 000	LEV160	0. 160	4. 2	4	0. 01
		ULEV125	0. 125	2. 1	4	0. 01
		ULEV70	0. 070	1. 7	4	0. 01
		ULEV50	0. 050	1. 7	4	0. 01
		SULEV30	0. 030	1. 0	4	0. 01
		SULEV20	0. 020	1. 0	4	0. 01

续表

Vehicle Type	Durability Vehicle Basis (*mi*)	Vehicle Emission Category[2]	NMOG + Oxides of Nitrogen (*g/mi*)	Carbon Monoxide (*g/mi*)	Formal-dehyde (*mg/mi*)	Particulates[1] (*g/mi*)
MDVs 8 501 – 10 000 lbs. GVWR, Vehicles in this category are tested at their adjusted loaded vehicle weight	150 000	LEV395	0. 395	6. 4	6	0. 12
		ULEV340	0. 340	6. 4	6	0. 06
		ULEV250	0. 250	6. 4	6	0. 06
		ULEV200	0. 200	4. 2	6	0. 06
		SULEV170	0. 170	4. 2	6	0. 06
		SULEV150	0. 150	3. 2	6	0. 06
MDVs 10 001 – 14 000 lbs. GVWR Vehicles in this category are tested at their adjusted loaded vehicle weight	150 000	LEV630	0. 630	7. 3	6	0. 12
		ULEV570	0. 570	7. 3	6	0. 06
		ULEV400	0. 400	7. 3	6	0. 06
		ULEV270	0. 270	4. 2	6	0. 06
		SULEV230	0. 230	4. 2	6	0. 06
		SULEV200	0. 200	3. 7	6	0. 06

[1] These standards shall apply only to vehicles not included in the phase-in of the particulate standards set forth in subsection (a)(2).

[2] The numeric portion of the category name is the NMOG + NOx value in thousandths of grams per mile.

(2) *"LEV III" Particulate Standards.*

(A) *Particulate Standards for Passenger Cars, Light-Duty Trucks, and Medium-Duty Passenger Vehicles.* Beginning in the 2017 model year, a manufacturer, except a small volume manufacturer, shall certify a percentage of its passenger car, light-duty truck, and medium-duty passenger vehicle fleet to the following particulate standards according to the following phase-in schedule. These standards are the maximum particulate emissions allowed at full useful life. All vehicles certifying to these particulate standards must certify to the LEV III exhaust emission standards set forth in subsection (a)(1).

LEV III Particulate Emission Standard Values and Phase-in for Passenger Cars, Light-Duty Trucks, and Medium-Duty Passenger Vehicles		
Model Year	*% of vehicles certified to a 3 mg/mi standard*	*% of vehicles certified to a 1 mg/mi standard*
2017	10	0
2018	20	0
2019	40	0
2020	70	0
2021	100	0
2022	100	0
2023	100	0
2024	100	0
2025	75	25
2026	50	50
2027	25	75
2028 and subsequent	0	100

（**B**）*Particulate Standards for Medium-Duty Vehicles Other than Medium-Duty Passenger Vehicles.*

1. Beginning in the 2017 model year, a manufacturer, except a small volume manufacturer, shall certify a percentage of its medium-duty vehicle fleet to the following particulate standards. These standards are the maximum particulate emissions allowed at full useful life. All vehicles certifying to these particulate standards must certify to the LEV III exhaust emission standards set forth in subsection（a）（1）. This subsection（a）（2）（B）1 shall not apply to medium-duty passenger vehicles.

LEV III Particulate Emission Standard Values for Medium-Duty Vehicles, Other than Medium-Duty Passenger Vehicles	
Vehicle Type[1]	*Particulates（mg/mi）*
MDVs 8 501 – 10 000 lbs. GVWR, excluding MDPVs	8
MDVs 10 001 – 14 000 lbs. GVWR	10

[1] Vehicles in these categories are tested at their adjusted loaded vehicle weight.

2. A manufacturer of medium-duty vehicles, except a small volume manufacturer, shall certify at least the following percentage of its medium-duty vehicle fleet to the particulate standards in subsection (a)(2)(B) 1 according to the following phase-in schedule. This subsection (a)(2)(B) 2 shall not apply to medium-duty passenger vehicles.

LEV III Particulate Emission Standard Phase-in for Medium-Duty Vehicles, Other than Medium-Duty Passenger Vehicles	
Model Year	*Total % of MDVs certified to the 8 mg/mi PM Standard or to the 10 mg/mi PM Standard, as applicable*
2017	10
2018	20
2019	40
2020	70
2021 and subsequent	100

(C) *Particulate Standards for Small Volume Manufacturers.* In the 2021 through 2027 model years, a small volume manufacturer shall certify 100 percent of its passenger car, light-duty truck, and medium-duty passenger vehicle fleet to the 3 mg/mi particulate standard. In the 2028 and subsequent model years, a small volume manufacturer shall certify 100 percent of its passenger car, light-duty truck, and medium-duty passenger vehicle fleet to the 1 mg/mi particulate standard. In the 2021 and subsequent model years, a small volume manufacturer shall certify 100 percent of its medium-duty vehicles 8 501 – 10 000 lbs. GVWR, excluding MDPVs, to the 8 mg/mi particulate standard. In the 2021 and subsequent model years, a small volume manufacturer shall certify 100 percent of its medium-duty vehicles 10 001 – 14 000 lbs. GVWR to the 10 mg/mi particulate standard. These standards are the maximum particulate emissions allowed at full useful life. All vehicles certifying to these particulate standards must certify to the LEV III exhaust emission standards set forth in subsection (a)(1).

(D) *Alternative Phase-in Schedule for Particulate Standards.*

1. *Alternative Phase-in Schedules for the 3 mg/mi Particulate Standard for Passenger Cars, Light-Duty Trucks, and Medium-Duty Passenger Vehicles.* A manufacturer may use an alternative phase-in schedule to comply with the 3 mg/mi particulate standard phase-in requirements as long as equivalent PM emission reductions are achieved by the 2021 model year from passenger cars, light-duty trucks, and medium-duty passenger vehicles. Model year emission reductions shall be calculated by multiplying the percent of PC + LDT + MDPV vehicles meeting the 3 mg/mi particulate standard in a given model year (based on a manufacturer's projected sales volume of vehicles in each category) by 5 for the 2017 model year, 4 for the 2018 model year, 3 for the 2019 model year, 2 for the 2020 model year, and 1 for the 2021 model year. The yearly results for PC + LDT + MDPV vehicles shall be summed together to determine a cumulative total for PC + LDT + MDPV vehicles. In the 2021 model year, the cumulative total must be equal to or greater than 490, and 100 percent of the manufacturer's passenger cars, light-duty trucks, and medium-duty passenger vehicles must be certified to the 3 mg/mi particulate standard to be considered equivalent. A manufacturer may add vehicles introduced before the 2017 model year (e. g., the percent of vehicles introduced in 2016 would be multiplied by 5) to the cumulative total.

2. *Alternative Phase-in Schedules for the 1 mg/mi Particulate Standard for Passenger Cars, Light-Duty Trucks, and Medium-Duty Passenger Vehicles*. A manufacturer may use an alternative phase-in schedule to comply with the 1 mg/mi particulate standard phase-in requirements as long as equivalent PM emission reductions are achieved by the 2028 model year from passenger cars, light-duty trucks, and medium-duty passenger vehicles. Model year emission reductions shall be calculated by multiplying the percent of PC + LDT + MDPV vehicles meeting the 1 mg/mi particulate standard in a given model year (based on a manufacturer's projected sales volume of vehicles in each category) by 4 for the 2025 model year, 3 for the 2026 model year, 2 for the 2027 model year, and 1 for the 2028 model year. The yearly results for PC + LDT + MDPV vehicles shall be summed together to determine a cumulative total for PC + LDT + MDPV vehicles. In the 2028 model year, the cumulative total must be equal to or greater than 500, and 100 percent of the manufacturer's passenger cars, light-duty trucks, and medium-duty passenger vehicles must be certified to the 1 mg/mi particulate standard to be considered equivalent. A manufacturer may add vehicles introduced before the 2025 model year (e. g., the percent of vehicles introduced in 2024 would be multiplied by 4) to the cumulative total.

3. *Alternative Phase-in Schedules for the Particulate Standards for Medium-Duty Vehicles Other than Medium-Duty Passenger Vehicles*. A manufacturer may use an alternative phase-in schedule to comply with the particulate standard phase-in requirements

as long as equivalent PM emission reductions are achieved by the 2021 model year from medium-duty vehicles other than medium-duty passenger vehicles. Model year emission reductions shall be calculated by multiplying the total percent of MDVs certified to the 8 mg/mi
PM standard or to the 10 mg/mi PM standard, as applicable, in a given model year (based
on a manufacturer's projected sales volume of vehicles in each category) by 5 for the 2017
model year, 4 for the 2018 model year, 3 for the 2019 model year, 2 for the 2020 model
year, and 1 for the 2021 model year. The yearly results for MDVs shall be summed together
to determine a cumulative total for MDVs. In the 2021 model year, the cumulative total
must be equal to or greater than 490, and 100 percent of the manufacturer's MDVs must be
certified to the 8 mg/mi PM standard or to the 10 mg/mi PM standard, as applicable, to be
considered equivalent. A manufacturer may add vehicles introduced before the 2017 model
year (e. g. , the percent of vehicles introduced in 2016 would be multiplied by 5) to the
cumulative total.

**(3) *NMOG + NOx Standards for Bi-Fuel, Fuel-Flexible, and Dual-Fuel Vehi-
cles.*** For fuel-flexible, bi-fuel, and dual-fuel PCs, LDTs and MDVs, compliance with the
NMOG + NOx exhaust mass emission standards must be based on exhaust emission tests both
when the vehicle is operated on the gaseous or alcohol fuel it is designed to use, and when
the vehicle is operated on gasoline. A manufacturer must demonstrate compliance with the
applicable exhaust mass emission standards for NMOG + NOx, CO, and formaldehyde set
forth in the table in subsection (a)(1) when certifying the vehicle for operation on the
gaseous or alcohol fuel, as applicable, and on gasoline or diesel, as applicable.

A manufacturer may measure NMHC in lieu of NMOG when fuel-flexible, bi-fuel and
dual-fuel vehicles are operated on gasoline, in accordance with the "California 2015 and
Subsequent Model Criteria Pollutant Exhaust Emission Standards and Test Procedures and
2017 and Subsequent Model Greenhouse Gas Exhaust Emission Standards and Test Procedures for Passenger Cars, Light-Duty Trucks, and Medium-Duty Vehicles. " Testing at 50°F
is not required for fuel-flexible, bi-fuel, and dual-fuel vehicles when operating on gasoline.

(4) *50°F Exhaust Emission Standards* . All passenger cars, light-duty trucks, and
medium-duty vehicles, other than natural gas and diesel-fueled vehicles, must demonstrate
compliance with the following exhaust emission standards for NMOG + NOx and formaldehyde (HCHO) measured on the FTP (40 CFR, Part 86, Subpart B) conducted at a
nominal test temperature of 50°F, as modified by Part II, Section C of the "California 2015
and Subsequent Model Criteria Pollutant Exhaust Emission Standards and Test Procedures

and 2017 and Subsequent Model Greenhouse Gas Exhaust Emission Standards and Test Procedures for Passenger Cars, Light-Duty Trucks, and Medium-Duty Vehicles. " A manufacturer may demonstrate compliance with the NMOG + NOx and HCHO certification standards contained in this subparagraph by measuring NMHC exhaust emissions or issuing a statement of compliance for HCHO in accordance with Section D. 1, subparagraph (p) and Section G. 3. 1. 2, respectively, of the "California 2015 and Subsequent Model Criteria Pollutant Exhaust Emission Standards and Test Procedures and 2017 and Subsequent Model Greenhouse Gas Exhaust Emission Standards and Test Procedures for Passenger Cars, Light-Duty Trucks, and Medium-Duty Vehicles. " Emissions of CO measured at 50°F shall not exceed the standards set forth in subsection (a)(1) applicable to vehicles of the same emission category and vehicle type subject to a cold soak and emission test at 68° to 86°F.

(A) *Standards for Passenger Cars, Light-Duty Trucks, and Medium-Duty Passenger Vehicles Certified to the LEV III Standards* .

50°F Exhaust Emission Standards for LEV III Passenger Cars, Light-Duty Trucks, and Medium-Duty Passenger Vehicles			
Vehicle Emission Category	NMOG + NOx (g/mi)		HCHO (g/mi)
	Gasoline	Alcohol Fuel	Both Gasoline and Alcohol Fuel
LEV160	0. 320	0. 320	0. 030
ULEV125	0. 250	0. 250	0. 016
ULEV70	0. 140	0. 250	0. 016
ULEV50	0. 100	0. 140	0. 016
SULEV30	0. 060	0. 125	0. 008
SULEV20	0. 040	0. 075	0. 008

(B) *Standards for Medium-Duty Vehicles (Excluding MDPVs) Certified to the LEV III Standards* .

50°F Exhaust Emission Standards for LEV III Medium-Duty Vehicles (Excluding MDPVs)			
Vehicle Emission Category	NMOG + NOx (g/mi)		HCHO (g/mi)
	Gasoline	Alcohol Fuel	Both Gasoline and Alcohol Fuel
LEV395	0. 790	0. 790	0. 064
ULEV340	0. 680	0. 680	0. 032
ULEV250	0. 500	0. 500	0. 032
ULEV200	0. 400	0. 500	0. 016
SULEV170	0. 340	0. 425	0. 016
SULEV150	0. 300	0. 375	0. 016
LEV630	1. 260	1. 260	0. 080
ULEV570	1. 140	1. 140	0. 042
ULEV400	0. 800	0. 800	0. 042
ULEV270	0. 540	0. 675	0. 020
SULEV230	0. 460	0. 575	0. 020
SULEV200	0. 400	0. 500	0. 020

(5) **Cold CO Standard** . The following standards are the 50 000 mile cold temperature exhaust carbon monoxide emission levels from new 2015 and subsequent model-year passenger cars, light-duty trucks, and medium-duty passenger vehicles:

2015 AND SUBSEQUENT MODEL – YEAR COLD TEMPERATURE CARBON MONOXIDE EXHAUST EMISSIONS STANDARDS FOR PASSENGER CARS, LIGHT – DUTY TRUCKS, AND MEDIUM – DUTY PASSENGER VEHICLES (grams per mile)	
Vehicle Type	Carbon Monoxide
All PCs, LDTs 0 – 3 750 lbs. LVW;	10. 0
LDTs, 3 751 lbs. LVW – 8 500 lbs. GVWR; MDPVs 10 000 lbs. GVWR and less	12. 5

These standards apply to vehicles tested at a nominal temperature of 20°F (−7℃) in accordance with 40 CFR Part 86 Subpart C, as amended by the "California 2015 and Subsequent Model Criteria Pollutant Exhaust Emission Standards and Test Procedures and 2017 and Subsequent Model Greenhouse Gas Exhaust Emission Standards and Test Procedures for

Passenger Cars, Light-Duty Trucks, and Medium-Duty Vehicles. " Natural gas, diesel-fueled and zero-emission vehicles are exempt from these standards.

(6) **Highway NMOG + NOx Standard** . The maximum emissions of non-methane organic gas plus oxides of nitrogen measured on the federal Highway Fuel Economy Test (HWFET; 40 CFR 600 Subpart B), as modified by the "California 2015 and Subsequent Model Criteria Pollutant Exhaust Emission Standards and Test Procedures and 2017 and Subsequent Model Greenhouse Gas Exhaust Emission Standards and Test Procedures for Passenger Cars, Light-Duty Trucks, and Medium-Duty Vehicles," must not be greater than the applicable LEV III NMOG + NOx standard set forth in subsection (a)(1). Both the sum of the NMOG + NOx emissions and the HWFET standard must be rounded in accordance with ASTM E29 – 67 to the nearest 0. 001 g/mi before being compared.

(7) **Supplemental Federal Test Procedure (SFTP) Off-Cycle Emission Standards** .

(A) **SFTP NMOG + NOx and CO Exhaust Emission Standards for Passenger Cars, Light-Duty Trucks, and Medium-Duty Passenger Vehicles.** Manufacturers shall certify 2015 and subsequent model year LEVs, ULEVs, and SULEVs in the PC, LDT, and MDPV classes to either the *SFTP NMOG + NOx and CO Stand-Alone Exhaust Emission Standards* set forth in subsection (a)(7)(A) 1, or in accordance with the *SFTP NMOG + NOx and CO Composite Exhaust Emission Standards and Fleet-Average Requirements* set forth in subsection (a)(7)(A) 2. A manufacturer may also certify 2014 model LEVs, ULEVs, or SULEVs in the PC, LDT, or MDPV classes to LEV III SFTP standards, in which case, the manufacturer shall be subject to the LEV III SFTP emission standards and requirements, including the sales-weighted fleet-average NMOG + NOx composite emission standard applicable to 2015 model vehicles if choosing to comply with the *SFTP NMOG + NOx and CO Composite Exhaust Emission Standards and Fleet-Average Requirements* set forth in subsection (a)(7)(A) 2. The manufacturer shall notify the Executive Officer of its selected emission standard type in the Application for Certification of the first test group certifying to SFTP NMOG + NOx and CO emission standards on a 150 000 mile durability basis. Once an emission standard type for NMOG + NOx and CO is selected for a fleet, and the Executive Officer is notified of such selection, the selection must be kept through the 2025 model year for the entire fleet, which includes LEV II vehicles if selecting to comply with subsection (a)(7)(A) 2. The manufacturer may not change its selection until the 2026 model year. Test groups not certifying to the 150 000 – mile SFTP NMOG + NOx and

CO emission standards pursuant to this subsection (a)(7)(A) shall be subject to the 4 000 – mile SFTP NMOG + NOx and CO emission standards set forth in subsection 1960. 1 (r).

1. *SFTP NMOG + NOx and CO Exhaust Stand-Alone Emission Standards.*
The following standards are the maximum SFTP NMOG + NOx and CO exhaust emissions through full useful life from 2015 and subsequent model-year LEV III LEVs, ULEVs, and SULEVs when operating on the same gaseous or liquid fuel they use for FTP certification. In the case of fuel-flexible vehicles, SFTP compliance shall be demonstrated using the LEV III certification gasoline specified in Part II, Section A. 100. 3. 1. 2 of the "California 2015 and Subsequent Model Criteria Pollutant Exhaust Emission Standards and Test Procedures and 2017 and Subsequent Model Greenhouse Gas Exhaust Emission Standards and Test Procedures for Passenger Cars, Light-Duty Trucks, and Medium-Duty Vehicles. ".

SFTP NMOG + NOx and CO Stand – Alone Exhaust Emission Standards for 2015 and Subsequent Model LEV III Passenger Cars, Light – Duty Trucks, and Medium – Duty Passenger Vehicles						
Vehicle Type	*Durability Vehicle Basis (mi)*	*Vehicle Emission Category[1]*	*US06 Test (g/mi)*		*SC03 Test (g/mi)*	
			NMOG + NOx	*CO*	*NMOG + NOx*	*CO*
All PCs; LDTs 0 – 8 500 lbs. GVWR; and MDPVs Vehicles in these categories are tested at their loaded vehicle weight (curbweight plus 300 pounds).	150 000	LEV	0. 140	9. 6	0. 100	3. 2
		ULEV	0. 120	9. 6	0. 070	3. 2
		SULEV (Option A)[2]	0. 060	9. 6	0. 020	3. 2
		SULEV	0. 050	9. 6	0. 020	3. 2

[1] ***Vehicle Emission Category.*** Manufacturers must certify all vehicles, which are certifying to a LEV III FTP emission category on a 150 000 – mile durability basis, to the emission standards of the equivalent, or a more stringent, SFTP emission category set forth on this table. That is, all LEV III LEVs certified to 150 000 – mile FTP emission standards shall comply with the SFTP LEV emission standards in this table, all LEV III ULEVs certified to 150 000 – mile FTP emission standards shall comply with the SFTP ULEV emission standards in this table, and all LEV III SULEVs certified to 150 000 – mile FTP emission standards shall comply with the SFTP SULEV emission standards in this table.

[2] ***Optional SFTP SULEV Standards.*** A manufacturer may certify light-duty truck test groups from 6 001 to 8 500 lbs. GVWR and MDPV test groups to the SULEV, option A, emission standards set forth in this table for the 2015 through 2020 model year, only if the vehicles in the test group are equipped with a particulate filter and the manufacturer extends the particulate filter emission warranty mileage to 200 000 miles. Passenger cars and light-duty trucks 0 – 6 000 lbs. GVWR are not eligible for this option.

2. SFTP NMOG + NOx and CO Composite Exhaust Emission Standards. For the 2015 and subsequent model years, a manufacturer selecting this option must certify LEV II and LEV III LEVs, ULEVs, and SULEVs, such that the manufacturer's sales-weighted fleet-average NMOG + NOx composite emission value does not exceed the applicable NMOG + NOx composite emission standard set forth in the following table. In addition, the CO composite emission value of any LEV III test group shall not exceed the CO composite emission standard set forth in the following table. SFTP compliance shall be demonstrated using the same gaseous or liquid fuel used for FTP certification. In the case of fuel-flexible vehicles, SFTP compliance shall be demonstrated wsing the LEV III certification gasoline specified in Part II, Section A. 100. 3. 1. 2 of the "California 2015 and Subsequent Model Criteria Pollutant Exhaust Emission Standards and Test Procedures and 2017 and Subsequent Model Greenhouse Gas Exhaust Emission Standards and Test Procedures for Passenger Cars, Light-Duty Trucks, and Medium-Duty Vehicles."

For each test group subject to this subsection, manufacturers shall calculate a Composite Emission Value for NMOG + NOx and, for LEV III test groups, a separate Composite Emission Value for CO, using the following equation:

$$Composite\ Emission\ Value = 0.28 \times US06 + 0.37 \times SC03 + 0.35 \times FTP \quad [Eq.\ 1]$$

where

"US06" = the test group's NMOG + NOx or CO emission value, as applicable, determined through the US06 test;

"SC03" = the test group's NMOG + NOx or CO emission value, as applicable, determined through the SC03 test; and

"FTP" = the test group's NMOG + NOx or CO emission value, as applicable, determined through the FTP test.

If no vehicles in a test group have air conditioning units, the FTP cycle emission value can be used in place of the SC03 cycle emission value in Equation 1. To determine compliance with the SFTP NMOG + NOx composite emission standard applicable to the model year, manufacturers shall use a sales-weighted fleet average of the NMOG + NOx composite emission values of every applicable test group. The sales-weighted fleet average shall be calculated using a combination of carry-over and new certification SFTP composite emission values (converted to NMOG + NOx, as applicable). LEV II test groups will use their emission values in the fleet average calculation but will not be considered LEV III test groups. Compliance with the CO composite emission standard cannot be demonstrated through fleet averaging. The NMOG + NOx sales-weighted fleet-average composite emission value for the fleet and the CO composite emission value for each test group shall not exceed：

SFTP NMOG + NOx and CO Composite Emission Standards for 2015 and Subsequent Model Passenger Cars , Light-Duty Trucks , and Medium-Duty Passenger Vehicles (g/mi)[1]											
Model Year	*2015*	*2016*	*2017*	*2018*	*2019*	*2020*	*2021*	*2022*	*2023*	*2024*	*2025 +*
All PCs; LDTs 8 500 lbs. GVWR or less; and MDPVs[3] Vehicles in this category are tested at their loaded vehicle weight (curb weight plus 300 pounds) except LEV Ⅱ vehicles, which are subject to the test weights specified in § 1960. 1 (r) , title 13 , CCR.	*Sales-Weighted Fleet Average NMOG + NOx Composite Exhaust Emission Standards*[2,4,5,6]										
	0. 140	0. 110	0. 103	0. 097	0. 090	0. 083	0. 077	0. 070	0. 063	0. 057	0. 050
	CO Composite Exhaust Emission Standard[7]										
	4. 2										

[1] ***Mileage for Compliance***. All test groups certifying to LEV III FTP emission standards on a 150 000 – mile durability basis shall also certify to the SFTP on a 150 000 – mile durability basis, as tested in accordance with the " California 2015 and Subsequent Model Criteria Pollutant Exhaust Emission Standards and Test Procedures and 2017 and Subsequent Model Greenhouse Gas Exhaust Emission Standards and Test Procedures for Passenger Cars, Light-Duty Trucks, and Medium-Duty Vehicles. "

[2] ***Determining NMOG + NOx Composite Emission Values of LEV II Test Groups and Cleaner Federal Vehicles***. For test groups certified to LEV II FTP emission standards, SFTP emission values shall be converted to NMOG + NOx and projected out to 120 000 miles or 150 000 miles (depending on LEV Ⅱ FTP artification) using deterioration factors or aged components. In lieu of deriving a deterioration factor specific to SFTP test cycles, carry-over LEV II test groups may use the applicable deterioration factor from the FTP cycle in order to determine the carry-over composite emission values for the purpose of the NMOG + NOx sales-weighted fleet-average calculation. If an SFTP full-useful life emission value is used to comply with the LEV II SFTP 4k standards, that value may be used in the sales-weighted fleet-average without applying an additional deterioration factor. For federally-certified test groups certifying in California in accordance with Section H. 1. 4 of the " California 2015 and Subsequent Model Criteria Pollutant Exhaust Emission

Standards and Test Procedures and 2017 and Subsequent Model Greenhouse Gas Exhaust Emission Standards and Test Procedures for Passenger Cars, Light-Duty Trucks, and Medium-Duty Vehicles," the full-useful life emission value used to comply with federal full-useful life SFTP requirements may be used in the sales-weighted fleet-average without applying an additional deterioration factor. In all cases, NMHC emission values for the US06 and SC03 test cycles shall be converted to NMOG emission values by multiplying by a factor of 1. 03.

[3] MDPVs are excluded from SFTP NMOG + NOx and CO emission standards and the sales-weighted fleet average until they are certified to LEV III FTP 150 000-mile NMOG + NOx and CO requirements.

[4] LEV III test groups shall certify to bins in increments of 0. 010 g/mi. Beginning with the 2018 model year, vehicles may not certify to bin values above a maximum of 0. 180 g/mi.

[5] *Calculating the sales-weighted average for NMOG + NOx.* For each model year, the manufacturer shall calculate its sales – weighted fleet – average NMOG + NOx composite emission value as follows.

$$\frac{\left[\sum_{i=1}^{n} (number\ of\ vehicles\ in\ the\ test\ group)_i \times (composite\ value\ of\ bin)_i \right]}{\sum_{i=1}^{n} (number\ of\ vehicles\ in\ the\ test\ group)_i} \quad [Eq.\ 2]$$

where "n" = a manufacturer's total number of PC, LDT, and, if applicable, MDPV certification bins, in a given model year including carry – over certification bins, certifying to SFTP composite emission standards in that model year;

"number of vehicles in the test group" = the number of vehicles produced and delivered for sale in California in the certification test group; and

"Composite Value of Bin" = the numerical value selected by the manufacturer for the certification bin that serves as the emission standard for the vehicles in the test group with respect to all testing for test groups certifying to SFTP on a 150 000 – mile durability basis, and the SFTP carry-over composite emission value, as described in footnote 2 of this table, for carry-over LEV II test groups.

[6] *Calculation of Fleet Average Total NMOG + NOx Credits or Debits*. A manufacturer shall calculate the total NMOG + NOx credits or debits, as follows:

[(NMOG + NOx Composite Emission Standard) – (Manufacturer's Sales – Weighted Fleet – Average Composite Emission Value)] × (Total Number of Vehicles Produced and Delivered for Sale in California in the 0 – 8 500 lbs GVWR plus MDPVs classes, if applicable) [Eq. 3]

A negative number constitutes total NMOG + NOx debits, and a positive number constitutes total NMOG + NOx credits accrued by the manufacturer for the given model year. Total NMOG + NOx credits earned in a given model year retain full value through the fifth model year after they are earned. At the beginning of the sixth model year, the total NMOG + NOx credits have no value. A manufacturer may trade credits with other manufacturers.

A manuafacturer shall equalize total NMOG + NOx debits within three model years after they have been incurred by earning NMOG + NOx credits in an amount equal to the total NMOG + NOx debits. If total NMOG + NOx debits are not equalized within the three model-year period, the manufacturer is subject to the Health and Safety Code section 43211 civil penalty applicable to a manufacturer which sells a new motor vehicle that does not meet the applicable emission standards adopted by the state board. The cause of action shall be deemed to accrue when the total NMOG + NOx debits are not equalized by the end of the specified time period. For the purposes of Health and Safety Code section 43 211, the number of vehicles not meeting the state board's emission standards is determined by dividing the NMOG + NOx debits for the model year by the NMOG + NOx composite emission standard in effect during the model year in which the debits were incurred.

[7] *Calculating the CO composite emission value.* Composite emission values for CO shall be calculated in accordance with Equation 1 above. Unlike the NMOG + NOx composite emission standards, manufacturers would not be able to meet the proposed CO composite emission standard through fleet averaging; each individual test group must comply with the standard. Test groups certified to 4 000 – mile SFTP emission

standards are not subject to this CO emission standard.

(**B**) *SFTP PM Exhaust Emission Standards for Passenger Cars*, *Light-Duty Trucks*, *and Medium-Duty Passenger Vehicles*. The following standards are the maximum PM exhaust emissions through the full useful life from 2017 and subsequent model-year LEV Ⅲ LEVs, ULEVs, and SULEVs in the PC, LDT, and MDPV classes when operating on the same gaseous or liquid fuel they use for FTP certification. In the case of fuel-flexible vehicles, SFTP compliance shall be demonstrated using the LEV Ⅲ certification gasoline specified in Part Ⅱ, Section A. 100. 3. 1. 2 of the "California 2015 and Subsequent Model Criteria Pollutant Exhaust Emission Standards and Test Procedures and 2017 and Subsequent Model Greenhouse Gas Exhaust Emission Standards and Test Procedures for Passenger Cars, Light-Duty Trucks, and Medium-Duty Vehicles." Manufacturers must certify LEVs, ULEVs, and SULEVs in the PC, LDT, and MDPV classes, which are certifying to LEV Ⅲ FTP PM emission standards in subsection (a)(2) on a 150 000 – mile durability basis, to the *SFTP PM Exhaust Emission Standards* set forth in this subsection (a)(7)(B).

SFTP PM Exhaust Emission Standards for 2017 and Subsequent Model LEV Ⅲ Passenger Cars, Light-Duty Trucks, and Medium-Duty Passenger Vehicles[1]				
Vehicle Type	*Test Weight*	*Mileage for Compliance*	*Test Cycle*	*PM (mg/mi)*
All PCs; LDTs 0 – 6 000 lbs GVWR	Loaded vehicle weight	150 000	US06	10
LDTs 6 001 – 8 500 lbs GVWR; MDPVs	Loaded vehicle weight	150 000	US06	20

[1] All PCs, LDTs, and MDPVs certified to LEV Ⅲ FTP PM emission standards in subsection (a) (2) on a 150, 000 – mile durability basis shall comply with the SFTP PM Exhaust Emission Standards in this table.

(**C**) *SFTP NMOG + NOx and CO Exhaust Emission Standards for Medium-Duty Vehicles*. The following standards are the maximum NMOG + NOx and CO composite emission values for full useful life of 2016 and subsequent model-year medium-duty LEV Ⅲ ULEVs and SULEVs from 8 501 through 14 000 pounds GVWR when operating on the same gaseous or liquid fuel they use for FTP certification. In the case of flex-fueled vehicles, SFTP compliance shall be demonstrated using the LEV Ⅲ certification gasoline specified in Part Ⅱ, Section A. 100. 3. 1. 2 of the "California 2015 and Subsequent Model Criteria Pollutant Exhaust Emission Standards and Test Procedures and 2017 and Subsequent Model Greenhouse Gas Exhaust Emission Standards and Test Procedures for Passenger Cars, Light-Duty Trucks, and Medium-Duty Vehicles. " The following composite emission standards do not apply to MDPVs subject to the emission standards presented in subsections (a)(7)(A) and (a)(7)(B).

SFTP NMOG + NOx and CO Composite Exhaust Emission Standards for 2016 and Subsequent Model ULEVs and SULEVs in the Medium-Duty Vehicle Class						
Vehicle Type	Mileage for Compliance	HP/ GVWR^2	Test Cycle^{3,4,5}	Vehicle Emission Category^6	Composite Emission Standard^1 (g/mi)	
					NMOG + NOx	Carbon Monoxide
MDVs 8 501 – 10 000 lbs GVWR	150 000	≤ 0.024	US06 Bag 2, SC03, FTP	ULEV	0.550	22.0
				SULEV	0.350	12.0
		> 0.024	Full US06, SC03, FTP	ULEV	0.800	22.0
				SULEV	0.450	12.0
MDVs 10 001 – 14 000 lbs GVWR	150 000	n/a	Hot 1435 UC (Hot 1435 LA92), SC03, FTP	ULEV	0.550	6.0
				SULEV	0.350	4.0

^{1} Manufacturers shall use Equation 1 in subsection (a)(7)(A) 2 to calculate SFTP Composite Emission Values for each test group subject to the emission standards in this table. For MDVs 10 001 – 14 000 lbs. GVWR, the emission results from the UC test shall be used in place of results from the US06 test.

^{2} *Power to Weight Ratio.* If all vehicles in a test group have a power to weight ratio at or below a threshold of 0.024, they may opt to run the US06 Bag 2 in lieu of the full US06 cycle. The cutoff is determined by using a ratio of the engine's maximum rated horsepower, as established by the engine manufacturer in the vehicle's Application for Certification, to the vehicle's GVWR in pounds and does not include any horsepower contributed by electric motors in the case of hybrid electric or plug-in hybrid electric vehicles. Manufacturers may opt to test to the full cycle regardless of the calculated ratio; in such case, manufacturers shall meet the emission standards applicable to vehicles with power-to-weight ratios greater than 0.024.

^{3} *Test Weight.* Medium-duty vehicles are tested at their adjusted loaded vehicle weight (average of curb weight and GVWR).

^{4} *Road Speed Fan.* Manufacturers have the option to use a road speed modulated fan as specified in 40 – CFR § 86.107 – 96 (d)(1) instead of a fixed speed fan for MDV SFTP testing.

^{5} If a manufacturer provides an engineering evaluation for a test group showing that SC03 emissions are equivalent to or lower than FTP emissions, the FTP emission value may be used in place of the SC03 emission value when determining the composite emission value for that test group.

^{6} *Vehicle Emission Categories.* For MDVs 8 501 – 10 000 lbs. GVWR, for each model year, the percentage of MDVs certified to an SFTP emission category set forth in this section 1961. 2 shall be equal to or greater than the total percentage certified to the FTP ULEV250, ULEV200, SULEV170, and SULEV150 emission categories; of these vehicles, the percentage of MDVs certified to an SFTP SULEV emission category shall be equal to or greater than the total percentage certified to both the FTP SULEV170 and SULEV150 emission categories. For MDVs 10 001 – 14 000 lbs. GVWR, for each model year, the percentage of MDVs certified to an SFTP emission category set forth in this section 1961. 2 shall be equal to or greater than the total percentage certified to the FTP ULEV400, ULEV270, SULEV230, and SULEV200 emission categories; of these vehicles, the percentage of MDVs certified to an SFTP SULEV emission category shall be equal to or greater than the total percentage certified to both the FTP SULEV230 and SULEV200 emission categories.

(D) *SFTP PM Exhaust Emission Standards for Medium-Duty Vehicles*. The following standards are the maximum PM composite emission values for the full useful life of 2017 and subsequent model-year LEV III LEVs, ULEVs, and SULEVs when operating on the same gaseous or liquid fuel they use for FTP certification. In the case of fuel-flexible vehicles, SFTP compliance shall be demonstrated using the LEV III certification gasoline specified in Part II, Section A. 100. 3. 1. 2 of the "California 2015 and Subsequent Model Criteria Pollutant Exhaust Emission Standards and Test Procedures and 2017 and Subsequent Model Greenhouse Gas Exhaust Emission Standards and Test Procedures for Passenger Cars, Light-Duty Trucks, and Medium-Duty Vehicles. " The following composite emission standards do not apply to MDPVs subject to the emission standards set forth in subsections (a)(7)(A) and (a)(7)(B).

SFTP PM Exhaust Emission Standards for 2017 and Subsequent Model Medium-Duty Vehicles[1]					
Vehicle Type	*Test Weight*	*Mileage for Compliance*	*Hp/GVWR*[2]	*Test Cycle*[3,4]	PM (*mg/mi*)
MDVs 8 501 – 10 000 lbs GVWR	Adjusted loaded vehicle weight	150 000	$\leqslant 0.024$	US06 Bag 2	7
			> 0. 024	US06	10
MDVs 10 001 – 14 000 lbs GVWR	Adjusted loaded vehicle weight	150 000	n/a	Hot 1435 UC (Hot 1435 LA92)	7

[1] Except for MDPVs subject to the emission standards set forth in subsection (a)(7)(B), MDVs certified to 150 000 – mile FTP PM emission standards in subsection (a)(2) shall comply with the SFTP PM Exhaust Emission Standards in this table.

[2] *Power to Weight Ratio*. If all vehicles in a test group have a power to weight ratio at or below a threshold of 0. 024, they may opt to run the US06 Bag 2 in lieu of the full US06 cycle. The cutoff is deter-

mined by using a ratio of the engine's horsepower to the vehicle's GVWR in pounds and does not include any horsepower contributed by electric motors in the case of hybrid electric or plug-in hybrid electric vehicles. Manufacturers may opt to test to the full cycle regardless of the calculated ratio; in such case, manufacturers shall meet the emission standards applicable to vehicles with power-to-weight ratios greater than 0. 024.

[3] *Road Speed Fan*. Manufacturers have the option to use a road speed modulated fan as specified in 40 – CFR § 86. 107 – 96 (d)(1) instead of a fixed speed fan for MDV SFTP testing.

[4] Manufacturers shall use Equation 1 above to calculate SFTP Composite PM Emission Values for each test group subject to the emission standards in this table. For MDVs 8 501 – 10 000 lbs. GVWR certifying to the US06 Bag 2 PM emission standard, the emission results from the US06 Bag 2 test shall be used in place of results from the full US06 test. For MDVs 10 001 – 14 000 lbs. GVWR, the emission results from the UC test shall be used in place of results from the US06 test.

(8) *Interim In-Use Compliance Standards*.

(A) *LEV III NMOG + NOx Interim In-Use Compliance Standards*. The following interim in-use compliance standards shall apply for the first two model years that a test group is certified to LEV III standards.

1. *NMOG + NOx Interim In-Use Compliance Standards for Passenger Cars, Light-Duty Trucks, and Medium-Duty Passenger Vehicles*. For the 2015 through 2019 model years, these standards shall apply.

Emission Category	Durability Vehicle Basis (miles)	LEV III PCs, LDTs, and MDPVs NMOG + NOx (g/mi)
LEV160	150 000	n/a
ULEV125	150 000	n/a
ULEV70	150 000	0. 098
ULEV50	150 000	0. 070
SULEV30	150 000	0. 042[1]
SULEV20	150 000	0. 028[1]

[1] not applicable to test groups that receive PZEV credits

2. *NMOG + NOx Interim In-Use Compliance Standards for Medium-Duty Vehicles, Excluding Medium-Duty Passenger Vehicles*. For the 2015 through 2020 model years, these standards shall apply.

Emission Category	Durability Vehicle Basis (miles)	LEV III MDVs, (excluding MDPVs) 8 501 – 10 000 lbs. GVW	LEV III MDVs, 10 001 – 14 000 lbs. GVW
		NMOG + NOx (g/mi)	NMOG + NOx (g/mi)
LEV395	150 000	n/a	n/a
ULEV340	150 000	n/a	n/a
ULEV250	150 000	0. 370	n/a
ULEV200	150 000	0. 300	n/a
SULEV170	150 000	0. 250	n/a
SULEV150	150 000	0. 220	n/a
LEV630	150 000	n/a	n/a
ULEV570	150 000	n/a	n/a
ULEV400	150 000	n/a	0. 600
ULEV270	150 000	n/a	0. 400
SULEV230	150 000	n/a	0. 340
SULEV200	150 000	n/a	0. 300

(B) *LEV III Particulate Interim In-Use Compliance Standards.* The following interim in-use compliance standards shall apply for the first two model years that a test group is certified to the LEV III standards.

1. LEV III Particulate Interim In-Use Compliance Standards for Passenger Cars, Light-Duty Trucks, and Medium-Duty Passenger Vehicles . For the 2017 through 2020 model years, the interim in-use compliance standard for vehicles certifying to the 3 mg/mi particulate standard is 6 mg/mi. For the 2025 through 2028 model years, the interim in-use compliance standard for vehicles certifying to the 1 mg/mi particulate standard is 2 mg/mi.

2. LEV III Particulate Interim In-Use Compliance Standards for Medium-Duty Vehicles, excluding Medium-Duty Passenger Vehicles. For the 2017 through 2020 model years, the interim in-use compliance standard for vehicles certifying to the 8 mg/mi particulate standard shall be 16 mg/mi and the interim in-use compliance standard for vehicles certifying to the 10 mg/mi particulate standard shall be 20 mg/mi.

(C) *SFTP Interim In-Use Compliance Standards.*

1. Test groups certified prior to the 2020 model year may use an in-use compliance standard for NMOG + NO$_x$ for the first two model years that they are certified to new standards.

a. For light-duty vehicle test groups and medium-duty passenger vehicle test groups certifying to the standards in subsection (a)(7)(A) 1, in-use compliance emission standards for NMOG + NOx shall be 1.4 times the applicable certification standard.

b. For light-duty vehicle test groups and medium-duty passenger vehicle test groups certifying to the standards in subsection (a)(7)(A) 2, in-use compliance emission standards for NMOG + NOx shall be 1.4 times the Composite Value of the bin to which a test group is certified.

c. For medium-duty vehicle tests groups certifying to the standards in subsection (a)(7)(C), in-use compliance emission standards for NMOG + NOx shall be 1.4 times the applicable certification standard.

2. Test groups certified prior to the 2021 model year will be allowed an in-use compliance standard for PM for the first five model year that they are certified to the SFTP PM standard.

a. For light-duty vehicle test groups and medium-duty passenger vehicle test groups certifying to SFTP PM exhaust emission standards in subsection (a)(7)(B), in-use compliance emission standards for PM shall be 5.0 mg/mi higher than the applicable certification standard.

b. For medium-duty vehicle test groups certifying to SFTP PM Exhaust Emission Standards in subsection (a)(7)(D), in-use compliance emission standards for PM shall be 5.0 mg/mi higher than the applicable certification standard.

(9) *Requirement to Generate Additional NMOG + NOx Fleet Average Credit.* For a vehicle that is certified to the LEV III standards in subsection (a)(1), which does not generate a partial ZEV allocation according to the criteria set forth in section C.3 of the "California Exhaust Emission Standards and Test Procedures for 2009 through 2017 Model Zero-Emission Vehicles and Hybrid Electric Vehicles, in the Passenger Car, Light-Duty Truck and Medium-Duty Vehicle Classes" and the "California Exhaust Emission Standards

and Test Procedures for 2018 and Subsequent Model Zero-Emission Vehicles and Hybrid E-lectric Vehicles, in the Passenger Car, Light-Duty Truck and Medium-Duty Vehicle Classes," a manufacturer may subtract 5 mg/mi from the NMOG + NOx emission standards value set forth in subsection (b)(1)(B) 1. c when calculating the manufacturer's fleet average, provided that the manufacturer extends the performance and defects warranty period to 15 years or 150 000 miles, whichever occurs first, except that the time period is to be 10 years for a zero emission energy storage device (such as battery, ultracapacitor, or other e-lectric storage device).

(10) ***Requirement to Generate a Partial ZEV Allowance.*** For the 2015 through 2017 model years, a manufacturer that certifies to the LEV III SULEV30 or the LEV III SU-LEV20 standards may also generate a partial ZEV allocation according to the criteria set forth in section C. 3 of the "California Exhaust Emission Standards and Test Procedures for 2009 through 2017 Model Zero-Emission Vehicles and Hybrid Electric Vehicles, in the Passenger Car, Light-Duty Truck and Medium-Duty Vehicle Classes. "

(11) ***NMOG Credit for Direct Ozone Reduction Technology.*** A manufacturer that certifies vehicles equipped with direct ozone reduction technologies shall be eligible to receive NMOG credits that can be applied to the NMOG exhaust emissions of the vehicle when determining compliance with the standard. In order to receive credit, the manufacturer must submit the following information for each vehicle model for which it gets credit, including, but not limited to:

(A) a demonstration of the airflow rate through the direct ozone reduction device and the ozone-reducing efficiency of the device over the range of speeds encountered in the Unified Cycle Driving Schedule contained in Part II G. of the "California 2015 and Subsequent Model Criteria Pollutant Emission Standards and Test Procedures for and 2017 and Subsequent Model Greenhouse Gas Exhaust Emission Standards and Test Procedures for Passenger Cars, Light-Duty trucks and Medium-duty Vehicles", as adopted March 22, 2012;

(B) an evaluation of the durability of the device for the full useful life of the vehicle; and

(C) a description of the on-board diagnostic strategy for monitoring the performance of the device in-use.

Using the above information, the Executive Officer shall determine the value of the NMOG credit based on the calculated change in the one-hour peak ozone level using an approved airshed model.

（12） *When a Federally-Certified Vehicle Model is Required in California.*

（A） *General Requirement.* Whenever a manufacturer federally-certifies a 2015 or subsequent model-year passenger car, light-duty truck, or medium-duty vehicle model to the standards for a particular emissions bin that are more stringent than the standards for an applicable California emission category, the equivalent California model may only be certified to （i）the California standards for a vehicle emissions category that are at least as stringent as the standards for the corresponding federal emissions bin, or （ii）the exhaust emission standards to which the federal model is certified. However, where the federal exhaust emission standards for the particular emissions bin and the California standards for a vehicle emissions category are equally stringent, the California model may only be certified to either the California standards for that vehicle emissions category or more stringent California standards. The federal emission bins are those contained in Tables S04 – 1 and S04 – 2 of 40 CFR § 86.1811 – 04 （c）as adopted February 10, 2000. The criteria for applying this requirement are set forth in Part I. Section H. 1 of the "California 2015 and Subsequent Model Criteria Pollutant Exhaust Emission Standards and Test Procedures and 2017 and Subsequent Model Greenhouse Gas Exhaust Emission Standards and Test Procedures for Passenger Cars, Light-Duty Trucks, and Medium-Duty Vehicles."

（B） *Exception for clean fuel fleet vehicles.* Subsection （a）（12）（A）does not apply in the case of a federally-certified vehicle model that is only marketed to fleet operators for applications that are subject to clean fuel fleet requirements established pursuant to section 246 of the federal Clean Air Act （42 U. S. C. sec. 7586）. In addition, the Executive Officer shall exclude from the requirement a federally-certified vehicle model where the manufacturer demonstrates to the Executive Officer's reasonable satisfaction that the model will primarily be sold or leased to clean fuel fleet operators for such applications, and that other sales or leases of the model will be incidental to marketing to those clean fuel fleet operators.

（13） *Emission Standard for a Fuel-Fired Heater.* Whenever a manufacturer elects to utilize an on-board fuel-fired heater on any passenger car, light-duty truck or medium-duty vehicle, the fuel-fired heater must meet ULEV125 standards for passenger cars and light-duty trucks less than 8 500 pounds GVWR as set forth in subsection 1961（a）（1）. The exhaust emissions from the fuel-fired heater shall be determined in accordance with the "California Exhaust Emission Standards and Test Procedures for 2009 through 2017 Model Zero-

Emission Vehicles and Hybrid Electric Vehicles, in the Passenger Car, Light-Duty Truck and Medium-Duty Vehicle Classes" or the "California Exhaust Emission Standards and Test Procedures for 2018 and Subsequent Model Zero-Emission Vehicles and Hybrid Electric Vehicles, in the Passenger Car, Light-Duty Truck and Medium-Duty Vehicle Classes," as applicable. If the on-board fuel-fired heater is capable of operating at ambient temperatures above 40°F, the measured emission levels of the on-board fuel-fired heater shall be added to the emissions measured on the FTP (40 CFR, Part 86, Subpart B), as amended by the "California 2015 and Subsequent Model Criteria Pollutant Exhaust Emission Standards and Test Procedures and 2017 and Subsequent Model Greenhouse Gas Exhaust Emission Standards and Test Procedures for Passenger Cars, Light-Duty Trucks, and Medium-Duty Vehicles" to determine compliance with the exhaust emission standards in subsection (a)(1).

(b) *Emission Standards Phase-In Requirements for Manufacturers.*

(1) *Fleet Average NMOG + NOx Requirements for Passenger Cars, Light-Duty Trucks, and Medium-Duty Passenger Vehicles.*

(A) The fleet average non-methane organic gas plus oxides of nitrogen exhaust mass emission values from the passenger cars, light-duty trucks, and medium-duty passenger vehicles that are produced and delivered for sale in California each model year by a manufacturer other than a small volume manufacturer shall not exceed:

FLEET AVERAGE NON-METHANE ORGANIC GAS PLUS OXIDES OF NITROGEN EXHAUST MASS EMISSION REQUIREMENTS FOR PASSENGER CARS, LIGHT-DUTY TRUCKS, AND MEDIUM-DUTY PASSENGER VEHICLES (*150 000 mile Durability Vehicle Basis*)		
	Fleet Average NMOG + NOx (*grams per mile*)	
Model Year	*All PCs; LDTs 0 – 3750 lbs. LVW*	*LDTs 3 751 lbs. LVW – 8 500 lbs. GVWR; All MDPVs*
2014[1]	0.107	0.128
2015	0.100	0.119
2016	0.093	0.110
2017	0.086	0.101

Model Year	Fleet Average NMOG + NOx (grams per mile)	
	All PCs; LDTs 0 – 3750 lbs. LVW	LDTs 3 751 lbs. LVW – 8 500 lbs. GVWR; All MDPVs
2018	0. 079	0. 092
2019	0. 072	0. 083
2020	0. 065	0. 074
2021	0. 058	0. 065
2022	0. 051	0. 056
2023	0. 044	0. 047
2024	0. 037	0. 038
2025 +	0. 030	0. 030

[1] For the 2014 model year, a manufacturer may comply with the fleet average NMOG + NOx values in this table in lieu of complying with the NMOG fleet average values in subsection 1961 (a)(b)(1)(A). A manufacturer must either comply with the NMOG + NOx fleet average requirements for both its PC/LDT1 fleet and its LDT2/MDPV fleet or comply with the NMOG fleet average requirements for both its PC/LDT1 fleet and its LDT2 fleet. A manufacturer must calculate its fleet average NMOG + NOx values using the applicable full useful life standards.

1. A manufacturer that selects compliance Option 2 must provide to the Executive Officer separate values for the number of vehicles in each test group produced and delivered for sale in the District of Columbia and for each individual state within the average.

2. *PZEV Anti-Backsliding Requirement.* In the 2018 and subsequent model years, a manufacturer must produce and deliver for sale in California a minimum percentage of its passenger car and light-duty truck fleet that certifies to SULEV30 and SULEV20 standards. This minimum percentage must be equal to the average percentage of PZEVs produced and deliver for sale in California for that manufacturer for the 2015 through 2017 model year. A manufacturer may calculate this average percentage using the projected sales for these model years in lieu of actual sales. The percentage of a manufacturer's passenger car and light-duty truck fleet that certifies to SULEV30 and SULEV20 standards averaged across the applicable model year and the two previous model years shall be used to determine compliance with this requirement, beginning with the 2020 model year.

(B) *Calculation of Fleet Average NMOG + NOx Value.*

1. *Basic Calculation.*

a. Each manufacturer's PC and LDT1 fleet average NMOG + NOx value for the total number of PCs and LDT1s produced and delivered for sale in California shall be calculated as follows:

($\sum$ [Number of vehicles in a test group excluding off-vehicle charge capable hybrid electric vehicles x applicable emission standard] + $\sum$ [Number of off – vehicle charge capable hybrid electric vehicles in a test group × HEV NMOG + NOx contribution factor]) ÷ Total Number of PCs plus LDT1s Produced and Delivered for sale in California, Including ZEVs and HEVs

b. Each manufacturer's LDT2 and MDPV fleet average NMOG + NOx value for the total number of LDT2s and MDPVs produced and delivered for sale in California shall be calculated as follows:

($\sum$ [Number of vehicles in a test group excluding off-vehicle charge capable hybrid electric vehicles × applicable emission standard] + $\sum$ [Number of off – vehicle charge capable hybrid electric vehicles in a test group × HEV NMOG factor]) ÷ Total Number of LDT2s plus MDPVs Produced and Delivered for sale in California, Including ZEVs and HEVs

c. The applicable emission standards to be used in the above equations are as follows:

Model Year	Emission Category	Emission Standard Value[1] (g/mi)	
		All PCs; LDTs 0 – 3750 lbs. LVW	LDTs 3 751 – 5 750 lbs. LVW; All MDPVs
2015 and subsequent model year federally-certified vehicles	All	Sum of the full useful life NMOG and NOx Federal Emission Standards to which Vehicle is Certified	Sum of the full useful life NMOG and NOx Federal Emission Standards to which Vehicle is Certified
Model Year	*Emission Category*	*All PCs; LDTs 0 – 3750 lbs. LVW*	*LDTs 3 751 lbs. LVW – 8 500 lbs. GVWR; All MDPVs*

Model Year	Emission Category	Emission Standard Value[1] (g/mi)	
		All PCs; LDTs 0 –3750 lbs. LVW	LDTs 3 751 lbs. LVW –8 500 lbs. GVWR All MDPVS LVW; All MDPVs
2015 through 2019 model year vehicles certified to the "LEV II" standards in subsection 1961 (a)(1); 2015 and subsequent model year vehicles certified to the "LEV III" standards in subsection 1961. 2 (a)(1)	LEV II LEVs; LEV160s	0. 160	0. 160
	LEV II ULEVs; LEV125s	0. 125	0. 125
	ULEV70s	0. 070	0. 070
	ULEV50s	0. 050	0. 050
	LEV II SULEVs; SULEV30s	0. 030	0. 030
	SULEV20s	0. 020	0. 020
	LEV II LEVs; LEV395s	n/a	0. 395
	LEV II ULEVs	n/a	0. 343
	ULEV340s	n/a	0. 340
	ULEV250s	n/a	0. 250
	ULEV200s	n/a	0. 200
	SULEV170s	n/a	0. 170
	SULEV150s	n/a	0. 150

[1] For LEV III vehicle test groups that meet the extended emission warranty requirements in subsection (a)(9), the applicable emission standard value shall be the emission standard value set forth in this table minus 5 mg/mi.

2. NMOG + NOx Contribution Factor for Off-vehicle Charge Capable HEVs.

The HEV NMOG + NOx contribution factor for light-duty off-vehicle charge capable hybrid electric vehicles is calculated as follows:

LEV160 HEV Contribution Factor = 0. 160 – [(Zero-emission VMT Allowance) × 0. 035]

ULEV125 HEV Contribution Factor = 0. 125 – [(Zero-emission VMT Allowance) × 0. 055]

ULEV70 HEV Contribution Factor = 0. 070 – [(Zero-emission VMT Allowance) × 0. 020]

ULEV50 HEV Contribution Factor = 0. 050 – [(Zero-emission VMT Allowance) × 0. 020]

SULEV30 HEV Contribution Factor = 0. 030 – [(Zero-emission VMT Allowance) × 0. 010]

SULEV20 HEV Contribution Factor =0. 020 − [(Zero-emission VMT Allowance) ×0. 020]

Where the Zero-emission VMT Allowance for off-vehicle charge capable HEVs is determined in accordance with section C. 3 of the "California Exhaust Emission Standards and Test Procedures for 2009 through 2017 Model Zero-Emission Vehicles and Hybrid Electric Vehicles, in the Passenger Car, Light-Duty Truck and Medium-Duty Vehicle Classes" and the "California Exhaust Emission Standards and Test Procedures for 2018 and Subsequent Model Zero-Emission Vehicles and Hybrid Electric Vehicles, in the Passenger Car, Light-Duty Truck and Medium-Duty Vehicle Classes," as applicable, except that for the purposes of this subsection (b)(1)(B) 2, the maximum allowable Zero-emission VMT Allowance that may be used in these equations is 1. 0. This subsection (b) (1) (B) 2 shall only apply to off-vehicle charge capable HEVs certified to the LEV Ⅲ standards set forth in subsection(a)(1).

(C) *Phase-In Requirements for Small Volume Manufacturers.*

1. In the 2015 through 2021 model years, a small volume manufacturer shall not exceed a fleet average NMOG + NOx value of 0. 160 g/mi for PCs and LDTs from 0 – 3 750 lbs. LVW or 0. 160 g/mi for LDTs from 3 751 – 5 750 lbs. LVW calculated in accordance with subsection (b)(1)(B). In the 2022 through 2024 model years, a small volume manufacturer shall not exceed a fleet average NMOG + NOx value of 0. 125 g/mi for PCs and LDTs from 0 – 3 750 lbs. LVW or 0. 125 g/mi for LDTs from 3 751 lbs. LVW – 8 500 lbs. GVW and MDPVs calculated in accordance with subsection (b)(1)(B). In 2025 and subsequent model years, a small volume manufacturer shall not exceed a fleet average NMOG + NOx value of 0. 070 g/mi for PCs and LDTs from 0 – 3 750 lbs. LVW or 0. 070 g/mi for LDTs from 3 751 lbs. LVW – 8 500 lbs. GVW and MDPVs calculated in accordance with subsection (b) (1) (B). For the 2015 through 2021 model years, a small volume manufacturer may certify its vehicles to the LEV II exhaust standards in section 1961. All vehicles certified by a small volume manufacturer for the 2022 and subsequent model years must meet the LEV III exhaust standards in this section 1961. 2.

2. If a manufacturer's average California sales exceeds 4500 units of new PCs, LDTs, MDVs, heavy-duty vehicles, and heavy-duty engines based on the average number of vehicles sold for the three previous consecutive model years, the manufacturer shall no longer be treated as a small volume manufacturer. If this is the first time the manufacturer exceeds the 4500 unit sales limit, the manufacturer must comply with the fleet average requirements applicable to a large volume manufacturer, as specified in subsection (b)(1) (A) beginning with the fourth model year after the last of the three consecutive model years. If during this four year lead time period the manufacturer's sales drop below the 4500 unit sales limit and then increase again above the 4500 unit sales limit, the four year lead time period shall be calculated based on the first model year in which the manufacturer again exceeds the 4500 unit sales limit. Except as noted above-i. e. , if this is not the first time the manufacturer has exceeded the 4500 unit sales limit-the manufacturer shall comply with the fleet average requirements applicable to larger manufacturers as specified in subsection (b)(1)(A) beginning with the following model year after the last of the three consecutive model years.

3. If a manufacturer's average California sales fall below 4500 units of new PCs, LDTs, MDVs and heavy duty engines based on the average number of vehicles sold for the three previous consecutive model years, the manufacturer shall be treated as a small volume manufacturer and shall be subject to the requirements for small volume manufacturers beginning with the next model year.

(D) *Treatment of ZEVs*. ZEVs classified as LDTs (>3750 lbs. LVW) that have been counted toward the ZEV requirement for PCs and LDTs (0 – 3750 lbs. LVW) as specified in sections 1962. 1 and 1962. 2 shall be included as LDT1s in the calculation of a fleet average NMOG + NOx value.

(2) *LEV III Phase-In Requirement for Passenger Cars, Light-Duty Trucks, and Medium-Duty Passenger Vehicles*. For the 2015 and 2016 model years, the LEV II SU-LEV emission standards set forth in section 1961 (a)(1) that are applicable to PCs, LDTs, and MDPVs shall only apply to those PCs, LDT1s, LDT2s, and MDPVs that certify to SULEV emission standards using "carryover" of emission test data from a previous model year in accordance with U. S. EPA OMS Advisory Circular A/C No. 17F, issued November 16, 1982, and last amended January 21, 1988, incorporated herein by reference. Beginning in the 2017 model year, the LEV II SULEV emission standards set forth in section 1961 (a)(1) that are applicable to PCs, LDTs, and MDPVs shall only apply to those PCs, LDT1s, LDT2s, and MDPVs that receive partial ZEV allowances in accordance with the "California Exhaust Emission Standards and Test Procedures for 2009 through 2017 Model Zero-Emission Vehicles and Hybrid Electric Vehicles, in the Passenger Car, Light-Duty Truck and Medium-Duty Vehicle Classes. " A manufacturer, other than a small volume manufacturer, must certify 100 percent of its PC, LDT, and MDPV fleet to the LEV III standards in subsection (a)(1) in 2020 and subsequent model years. A small volume manufacturer must certify 100 percent of its PC, LDT, and MDPV fleet to the LEV III standards in subsection （a）（1） in 2022 and subsequent model years.

（3） *LEV III Phase-In Requirements for Medium-Duty Vehicles, Other than Medium-Duty Passenger Vehicles.*

（A） A manufacturer of MDVs, other than a small volume manufacturer, shall certify its MDV fleet according to the following phase-in schedule:

Model Year	Vehicles Certified to § 1961. 2 （a）（1） （%）				Vehicles Certified to § 1956. 8 （c） or （h）（%）
	LEV II LEV; LEV III LEV395 or LEV630	LEV II ULEV; LEV III ULEV340 or ULEV570	LEV III ULEV250 or ULEV400	LEV III SULEV170 or SULEV230	ULEV
2015	40	60	0	0	100
2016	20	60	20	0	100
2017	10	50	40	0	100
2018	0	40	50	10	100
2019	0	30	40	30	100
2020	0	20	30	50	100
2021	0	10	20	70	100
2022 +	0	0	10	90	100

（B） *Requirements for Small Volume Manufacturers.* In the 2015 through 2017 model years, a small volume manufacturer shall certify, produce, and deliver for sale in California vehicles or engines certified to the MDV LEV II LEV standards or to the LEV III LEV395 or LEV III LEV630 standards, as applicable, in a quantity equivalent to 100% of its MDV fleet. In the 2018 through 2021 model years, a small volume manufacturer shall certify, produce, and deliver for sale in California vehicles or engines certified to the MDV LEV II ULEV standards or to the LEV III ULEV340 or LEV III ULEV570 standards, as applicable, in a quantity equivalent to 100% of its MDV fleet. In the 2022 and subsequent model years, a small volume manufacturer shall certify, produce, and deliver for sale in California vehicles or engines certified to the MDV LEV III ULEV250 or LEV III ULEV400 standards, as applicable, in a quantity equivalent to 100% of its MDV fleet. Engines certified to these MDV standards are not eligible for emissions averaging.

（C） *Alternate Phase-In Schedules for LEV III MDVs.* For the 2016 and subsequent model years, a manufacturer that produces and delivers for sale in California four or fewer medium-duty test groups may comply with the following alternate phase-in schedule for LEV III medium-duty vehicles.

1. A manufacturer that produces and delivers for sale in California four medium-duty test groups may comply with the following alternate phase-in schedule for LEV III medium-duty vehicles.

Model Year	Number of Test Groups Certified to § 1961. 2 (a) (1)				Vehicles Certified to § 1956. 8 (c)or(h) (%)
	LEV II LEV; LEV III LEV395 or LEV630	LEV II ULEV; LEV III ULEV340 or ULEV570	LEV III; ULEV 250 or ULEV400	LEV III SULEV170 or SULEV230	ULEV
2016 – 2017	1	2	1	0	100
2018	0	2	2	0	100
2019	0	1	2	1	100
2020	0	1	1	2	100
2021	0	0	1	3	100
2022 +	0	0	0	4	100

2. A manufacturer that produces and delivers for sale in California three medium-duty test groups may comply with the following alternate phase-in schedule for LEV III medium-duty vehicles.

Model Year	Number of Test Groups Certified to § 1961. 2 (a) (1)				Vehicles Certified to § 1956. 8 (c)or(h) (%)
	LEV II LEV; LEV III LEV395 or LEV630	LEV II ULEV; LEV III ULEV340 or ULEV570	LEV III; ULEV 250 or ULEV400	LEV III SULEV170 or SULEV230	ULEV
2016	1	2	0	0	100
2017	0	2	1	0	100

续表

Model Year	Number of Test Groups Certified to § 1961. 2 (a) (1)				Vehicles Certified to § 1956. 8 (c) or (h) (%)
	LEV Ⅱ LEV; LEV Ⅲ LEV395 or LEV630	LEV Ⅱ ULEV; LEV Ⅲ ULEV340 or ULEV570	LEV Ⅲ; ULEV 250 or ULEV400	LEV Ⅲ SULEV170 or SULEV230	ULEV
2018	0	1	2	0	100
2019 – 2020	0	1	1	1	100
2021	0	0	1	2	100
2022 +	0	0	0	3	100

3. A manufacturer that produces and delivers for sale in California two medium-duty test groups may comply with the following alternate phase-in schedule for LEV Ⅲ medium-duty vehicles.

Model Year	Number of Test Groups Certified to § 1961. 2 (a) (1)				Vehicles Certified to § 1956. 8 (c) or (h) (%)
	LEV Ⅱ LEV; LEV Ⅲ LEV395 or LEV630	LEV Ⅱ ULEV; LEV Ⅲ ULEV340 or ULEV570	LEV Ⅲ; ULEV 250 or ULEV400	LEV Ⅲ SULEV170 or SULEV230	ULEV
2016	1	1	0	0	100
2017 – 2019	0	1	1	0	100
2020 – 2021	0	0	1	1	100
2022 +	0	0	0	2	100

4. A manufacturer that produces and delivers for sale in California one medium-duty test groups may comply with the following alternate phase-in schedule for LEV Ⅲ medium-duty vehicles.

Model Year	Number of Test Groups Certified to § 1961.2 (a) (1)				Vehicles Certified to § 1956.8 (c) or (h) (%)
	LEV II LEV; LEV III LEV395 or LEV630	LEV II ULEV; LEV III ULEV340 or ULEV570	LEV III; ULEV 250 or ULEV400	LEV III SULEV170 or SULEV230	ULEV
2016 – 2018	0	1	0	0	100
2019 – 2021	0	0	1	0	100
2022 +	0	0	0	1	100

(D) Identifying a Manufacturer's MDV Fleet. Each manufacturer's MDV fleet shall be defined as the total number of California-certified MDVs produced and delivered for sale in California. The percentages shall be applied to the manufacturer's total production of California-certified medium-duty vehicles delivered for sale in California. A manufacturer that elects to certify to the optional medium-duty engine standards in subsections 1956.8 (c) or (h) shall not count those engines in the manufacturer's total production of California-certified medium-duty vehicles for purposes of this subsection.

(E) For a manufacturer that elects to certify to the optional medium-duty engine standards in title 13, CCR subsections 1956.8 (c) or (h), all such MDVs, including those produced by a small volume manufacturer, shall be subject to the emissions averaging provisions applicable to heavy-duty diesel or Otto-cycle engines as set forth in the "California Exhaust Emission Standards and Test Procedures for 2004 and Subsequent Model Heavy-Duty Otto-Cycle Engines," or the "California Exhaust Emission Standards and Test Procedures for 2004 and Subsequent Model Heavy-Duty Diesel Engines, incorporated by reference in subsections 1956.8 (b) or (d), as applicable.

(4) SFTP Phase-In Requirements.

(A) Phase-In Requirement for Passenger Cars, Light-Duty Trucks, and Medium-Duty Passenger Vehicles. A test group certifying to LEV III FTP emission categories on a 150 000 – mile durability basis shall also certify to SFTP requirements on a 150 000 – mile durability basis.

Manufacturers shall have two options for phase in to the SFTP NMOG + NOx and CO emission standards.

1. Under Option 1, beginning with the 2015 model year, a manufacturer shall certify its PCs, LDTs, and MDPVs to the SFTP NMOG + NOx and CO emission standards in subsection (a)(7)(A) 1 when the vehicles are also certifying to a LEV III FTP emission category at 150 000 – mile durability.

2. Under Option 2, for 2015 and subsequent model years, a manufacturer shall certify its fleet of PCs, LDTs, and MDPVs such that the manufacturer's sales-weighted fleet-average NMOG + NOx composite emission value and each test group's CO composite emission value does not exceed the applicable composite emission standards in effect for that model year in accordance with subsection (a)(7)(A) 2.

Beginning with the 2017 model year, a manufacturer shall certify its PCs, LDTs, and MD-PVs certifying to LEV III FTP PM emission standards on a 150 000 – mile durability basis to the SFTP PM emission standards in subsection (a)(7)(B).

(B) *Phase-In Requirements for Medium-Duty Vehicle Manufacturers*. Phase-in for NMOG + NOx and CO emission standards begins with the 2016 model year. For MD-Vs 8 501 – 10 000 lbs. GVWR, for each model year, the percentage of MDVs certified to an SFTP emission category set forth in this section 1961. 2 shall be equal to or greater than the total percentage certified to the FTP ULEV250, ULEV200, SULEV170, and SU-LEV150 emission categories; of these vehicles, the percentage of MDVs certified to an SFTP SULEV emission category shall be equal to or greater than the total percentage certified to both the FTP SULEV170 and SULEV150 emission categories. For MDVs 10 001 – 14 000 lbs. GVWR, for each model year, the percentage of MDVs certified to an SFTP e-mission category set forth in this section 1961. 2 shall be equal to or greater than the total percentage certified to the FTP ULEV400, ULEV270, SULEV230, and SULEV200 emission categories; of these vehicles, the percentage of MDVs certified to an SFTP SULEV e-mission category shall be equal to or greater than the total percentage certified to both the FTP SULEV230 and SULEV200 emission categories.

In addition, 2017 and subsequent model MDVs certifying to LEV III FTP PM emission standards on a 150 000 – mile durability basis must also certify to the SFTP emission standards set forth in subsection (a)(7)(D).

(C) *Identifying a Manufacturer's Medium-Duty Vehicle Fleet*. For the 2016 and subsequent model years, each manufacturer's MDV fleet shall be defined as the total number of California-certified MDVs, other than MDPVs, produced and delivered for sale in California. For 2016 and subsequent model years, a manufacturer that elects to certify en-

gines to the optional medium-duty engine emission standards in subsections 1956. 8 （c） or
（h） shall not count those engines in the manufacturer's total production of California-certified medium-duty vehicles for purposes of this subparagraph.

（c） *Calculation of NMOG + NOx Credits/Debits*

（1） *Calculation of NMOG + NOx Credits and Debits for Passenger Cars, Light-Duty Trucks, and Medium-Duty Passenger Vehicles*.

（A） In 2015 and subsequent model years, a manufacturer shall calculate its
credits or debits using the following equation.

[（Fleet Average NMOG + NOx Requirement） – （Manufacturer's Fleet Average NMOG
+ NOx Value）] x

（Total No. of Vehicles Produced and Delivered for Sale in California, Including ZEVs
and HEVs）.

（B） In 2015 and subsequent model years, a manufacturer that achieves fleet average NMOG + NOx values lower than the fleet average NMOG + NOx requirement for the
corresponding model year shall receive credits in units of g/mi NMOG + NOx. A manufacturer with 2015 and subsequent model year fleet average NMOG + NOx values greater than
the fleet average requirement for the corresponding model year shall receive debits in units
of g/mi NMOG + NOx equal to the amount of negative credits determined by the aforementioned equation. The total g/mi NMOG + NOx credits or debits earned for PCs and LDTs 0
– 3750 lbs. LVW, and for LDTs 3 751 lbs. LVW – 8 500 lbs. GVWR and for MDPVs
shall be summed together. The resulting amount shall constitute the g/mi NMOG + NOx
credits or debits accrued by the manufacturer for the model year.

**（2） *Calculation of Vehicle-Equivalent NMOG + NOx Credits for Medium-Duty
Vehicles Other than MDPVs*.**

（A） In 2016 and subsequent model years, a manufacturer that produces and delivers for sale in California MDVs, other than MDPVs, in excess of the equivalent requirements for LEV III vehicles certified to the exhaust emission standards set forth in subsection
（a）（1）, shall receive "Vehicle-Equivalent Credits" （or "VECs"） calculated in accordance with the following equation, where the term "produced" means produced and delivered
for sale in California：

（1. 00） × {[（No. of LEV395s and LEV630s Produced excluding HEVs） +
（No. of LEV395 HEVs × HEV VEC factor for LEV395s） +
（No. of LEV630 HEVs × HEV VEC factor for LEV630s）] –

(No. of LEV395s and LEV630s Required to be Produced)} +

(1. 14) × {[(No. of ULEV340s and ULEV570s Produced excluding HEVs) +

(No. of ULEV340 HEVs × HEV VEC factor for ULEV340s) +

(No. of ULEV570 HEVs × HEV VEC factor for ULEV570s)] −

(No. of ULEV340s and ULEV570s Required to be Produced)} +

(1. 37) × {[(No. of ULEV250s and ULEV400s Produced excluding HEVs) +

(No. of ULEV250 HEVs × HEV VEC factor for ULEV250s) +

(No. of ULEV400 HEVs × HEV VEC factor for ULEV400s)] −

(No. of ULEV250s and ULEV400s Required to be Produced)} +

(1. 49) × {[(No. of ULEV200s and ULEV270s Produced excluding HEVs) +

(No. of ULEV200 HEVs × HEV VEC factor for ULEV200s) +

(No. of ULEV270 HEVs × HEV VEC factor for ULEV270s)] −

(No. of ULEV200s and ULEV270s Required to be Produced)} +

(1. 57) × {[(No. of SULEV170s and SULEV230s Produced excluding HEVs) +

(No. of SULEV170 HEVs × HEV VEC factor for SULEV170s) +

(No. of SULEV230 HEVs × HEV VEC factor for SULEV230s)] −

(No. of SULEV170s and SULEV230s Required to be Produced)} +

(1. 62) × {[(No. of SULEV150s and SULEV200s Produced excluding HEVs) +

(No. of SULEV150 HEVs × HEV VEC factor for SULEV150s) +

(No. of SULEV200 HEVs × HEV VEC factor for SULEV200s)] −

(No. of SULEV150s and SULEV200s Required to be Produced)} +

[(2. 00) × (No. of ZEVs Certified and Produced as MDVs)].

(B) *MDV HEV VEC factor*. The MDV HEV VEC factor is calculated as follows:

For LEV395s:

$$1 + \left[\frac{(\text{LEV395 standard-ULEV340 standard}) \times \text{Zero-emission VMT Allowance}}{\text{LEV395 standard}} \right];$$

For ULEV340s:

$$1 + \left[\frac{(\text{ULEV340 standard-ULEV250 standard}) \times \text{Zero-emission VMT Allowance}}{\text{ULEV340 standard}} \right];$$

For ULEV250s:

$$1 + \left[\frac{(\text{ULEV250 standard-ULEV200 standard}) \times \text{Zero-emission VMT Allowance}}{\text{ULEV250 standard}} \right];$$

For ULEV200s：

$$1 + \left[\frac{(\text{ULEV200 standard-SULEV170 standard}) \times \text{Zero-emission VMT Allowance}}{\text{ULEV200 standard}}\right];$$

For SULEV170s：

$$1 + \left[\frac{(\text{SULEV170 standard-SULEV150 standard}) \times \text{Zero-emission VMT Allowance}}{\text{SULEV170 standard}}\right];$$

For SULEV 150s：

$$1 + \left[\frac{(\text{SULEV150 standard-ZEV standard}) \times \text{Zero-emission VMT Allowance}}{\text{SULEV150 standard}}\right];$$

For LEV630s：

$$1 + \left[\frac{(\text{LEV630 standard-ULEV570 standard}) \times \text{Zero-emission VMT Allowance}}{\text{LEV630 standard}}\right];$$

For ULEV570s：

$$1 + \left[\frac{(\text{ULEV570 standard-ULEV400 standard}) \times \text{Zero-emission VMT Allowance}}{\text{ULEV570 standard}}\right];$$

For ULEV400s：

$$1 + \left[\frac{(\text{ULEV400 standard-ULEV270 standard}) \times \text{Zero-emission VMT Allowance}}{\text{ULEV400 standard}}\right];$$

For ULEV270s：

$$1 + \left[\frac{(\text{ULEV270 standard-SULEV230 standard}) \times \text{Zero-emission VMT Allowance}}{\text{ULEV270 standard}}\right];$$

ForSULEV230s：

$$1 + \left[\frac{(\text{SULEV230 standard-SULEV200 standard}) \times \text{Zero-emission VMT Allowance}}{\text{SULEV230 standard}}\right]$$

ForSULEV200s：

$$1 + \left[\frac{(\text{SULEV200 standard-ZEV standard}) \times \text{Zero-emission VMT Allowance}}{\text{SULEV200 standard}}\right]$$

Where "Zero-emission VMT Allowance" for an HEV is determined in accordance with section C of the "California Exhaust Emission Standards and Test Procedures for 2009 through 2017 Model Zero-Emission Vehicles and Hybrid Electric Vehicles, in the Passenger Car, Light-Duty Truck and Medium-Duty Vehicle Classes," incorporated by reference in section 1962. 1, or the "California Exhaust Emission Standards and Test Procedures for 2018 and Subsequent Model Zero-Emission Vehicles and Hybrid Electric Vehicles, in the Passenger Car, Light-Duty Truck and Medium-Duty Vehicle Classes," incorporated by reference in section 1962. 2, as applicable, except that for the purposes of this subsection （c）（2）（B）, the maximum allowable Zero-emission VMT Allowance that may be used in these equations is 1. 0.

(**C**) A manufacturer that fails to produce and deliver for sale in California the e-quivalent quantity of MDVs certified to LEV III exhaust emission standards, shall receive "Vehicle-Equivalent Debits" (or "VEDs") equal to the amount of negative VECs deter-mined by the equation in subsection (c)(2)(A).

(**D**) Only ZEVs certified as MDVs and not used to meet the ZEV requirement shall be included in the calculation of VECs.

(3) *Procedure for Offsetting Debits.*

(**A**) A manufacturer shall equalize emission debits by earning g/mi NMOG + NOx emission credits or VECs in an amount equal to the g/mi NMOG + NOx debits or VEDs, or by submitting a commensurate amount of g/mi NMOG + NOx credits or VECs to the Execu-tive Officer that were earned previously or acquired from another manufacturer. A manufac-turer shall equalize NMOG + NOx debits for PCs, LDTs, and MDPVs and VEC debits for MDVs within three model years. If emission debits are not equalized within the specified time period, the manufacturer shall be subject to the Health and Safety Code § 43211 civil penalty applicable to a manufacturer which sells a new motor vehicle that does not meet the applicable emission standards adopted by the state board. The cause of action shall be deemed to accrue when the emission debits are not equalized by the end of the specified time period. A manufacturer demonstrating compliance under Option 2 in subsection (b) (1)(A) 1. a, must calculate the emission debits that are subject to a civil penalty under Health and Safety Code section 43211 separately for California, the District of Columbia, and for each individual state that is included in the fleet average greenhouse gas require-ments in subsection (b)(1)(A) 1. a. The manufacturer must calculate these emission debits separately for California, the District of Columbia, and each individual state using the formula in subsections (c)(1) and (c)(2), except that the "Total No. of Vehicles Pro-duced and Delivered for Sale in California, Including ZEVs and HEVs" shall be calculated separately for the District of Columbia and each individual state.

For the purposes of Health and Safety Code § 43211, the number of passenger cars, light-duty trucks, and medium-duty passenger vehicles not meeting the state board's emission standards shall be determined by dividing the total amount of g/mi NMOG + NOx emission debits for the model year by the g/mi NMOG + NOx fleet average requirement for PCs and LDTs 0 – 3 750 lbs. LVW and for LDTs 3 751 lbs. LVW – 8 500 lbs. GVW and MDPVs applicable for the model year in which the debits were first incurred; and the number of medium-duty vehicles not meeting the state board's emission standards shall be equal to the amount of VEDs incurred.

（**B**）The emission credits earned in any given model year shall retain full value through five subsequent model years. Credits will have no value if not used by the beginning of the sixth model year after being earned.

（4）*Changing NMOG Credits and Debits to NMOG + NOx Credits and Debits*. The value of any emission credits that have not been used prior to the start of the 2015 model year and any emission debits that have not been equalized prior to the start of the 2015 model year earned shall be converted to NMOG + NOx credits at the start of the 2015 model year by multiplying their values by a factor of 3.0. These credits and debits are subject to the provisions in subsection 1961（c）（3）.

（**d**）*Test Procedures*. The certification requirements and test procedures for determining compliance with the emission standards in this section are set forth in the "California 2015 and Subsequent Model Criteria Pollutant Exhaust Emission Standards and Test Procedures and 2017 and Subsequent Model Greenhouse Gas Exhaust Emission Standards and Test Procedures for Passenger Cars, Light-Duty Trucks, and Medium-Duty Vehicles," as amended December 6, 2012, the "California Non-Methane Organic Gas Test Procedures," as amended December 6, 2012, which are incorporated herein by reference. In the case of hybrid electric vehicles and on-board fuel-fired heaters, the certification requirements and test procedures for determining compliance with the emission standards in this section are set forth in the "California Exhaust Emission Standards and Test Procedures for 2009 through 2017 Model Zero-Emission Vehicles and Hybrid Electric Vehicles, in the Passenger Car, Light-Duty Truck and Medium-Duty Vehicle Classes," incorporated by reference in section 1962.1, and the "California Exhaust Emission Standards and Test Procedures for 2018 and Subsequent Model Zero-Emission Vehicles and Hybrid Electric Vehicles, in the Passenger Car, Light-Duty Truck and Medium-Duty Vehicle Classes," incorporated by reference in section 1962.2.

（**e**）*Abbreviations*. The following abbreviations are used in this section 1961.2:

"ALVW" means adjusted loaded vehicle weight.

"ASTM" means American Society of Testing and Materials.

"CO" means carbon monoxide.

"FTP" means Federal Test Procedure.

"g/mi" means grams per mile.

"GVW" means gross vehicle weight.

"GVWR" means gross vehicle weight rating.

"HEV" means hybrid-electric vehicle.

"LDT" means light-duty truck.

"LDT1" means a light-duty truck with a loaded vehicle weight of 0 – 3750 pounds.

"LDT2" means a light-duty truck with a loaded vehicle weight of 3 751 pounds to a gross vehicle weight rating of 8 500 pounds.

"LEV" means low-emission vehicle.

"LPG" means liquefied petroleum gas.

"LVW" means loaded vehicle weight.

"MDPV" means medium-duty passenger vehicle.

"MDV" means medium-duty vehicle.

"NMHC" means non-methane hydrocarbons.

"mg/mi" means milligrams per mile.

"NMHC" means non-methane hydrocarbons.

"Non-Methane Organic Gases" or "NMOG" means the total mass of oxygenated and non-oxygenated hydrocarbon emissions.

"NOx" means oxides of nitrogen.

"PC" means passenger car.

"SULEV" means super-ultra-low-emission vehicle.

"ULEV" means ultra-low-emission vehicle.

"VEC" means vehicle-equivalent credits.

"VED" means vehicle-equivalent debits.

"VMT" means vehicle miles traveled.

"ZEV" means zero-emission vehicle.

(f) *Severability*. Each provision of this section is severable, and in the event that any provision of this section is held to be invalid, the remainder of both this section and this article remains in full force and effect.

Note: Authority cited: Sections 39500, 39600, 39601, 43013, 43018, 43101, 43104, 43105 and 43106, Health and Safety Code. Reference: Sections 39002, 39003, 39667, 43000, 43009.5, 43013, 43018, 43100, 43101, 43101.5, 43102, 43104, 43105, 43106, 43204 and 43205, Health and Safety Code.

§ 1961. 3 Greenhouse Gas Exhaust Emission Standards and Test Procedures – 2017 and Subsequent Model Passenger Cars, Light-Duty Trucks, and Medium-Duty Vehicles.

Introduction. This section 1961. 3 sets the greenhouse gas emission levels from new 2017 and subsequent model year passenger cars, light-duty trucks, and medium-duty passenger vehicles. Light-duty trucks from 3 751 lbs. LVW – 8 500 lbs. GVW that are certified to the Option 1 LEV II NOx Standard in section 1961 (a)(1) are exempt from these greenhouse gas emission requirements, however, passenger cars, light-duty trucks 0 – 3 750 lbs. LVW, and medium-duty passenger vehicles are not eligible for this exemption.

Emergency vehicles may be excluded from these greenhouse gas emission requirements. The manufacturer must notify the Executive Officer that they are making such an election, in writing, prior to the start of the applicable model year or must comply with this section 1961. 3.

(a) ***Greenhouse Gas Emission Requirements.***

(1) ***Fleet Average Carbon Dioxide Requirements for Passenger Cars, Light-Duty Trucks, and Medium-Duty Passenger Vehicles.*** For the purpose of determining compliance with this subsection (a)(1), the applicable fleet average CO_2 mass emission standards for each model year is the sales-weighted average of the calculated CO_2 exhaust mass emission target values for each manufacturer. For each model year, the sales-weighted fleet average CO_2 mass emissions value shall not exceed the sales-weighted average of the calculated CO_2 exhaust mass emission target values for that manufacturer.

(A) ***Fleet Average Carbon Dioxide Target Values for Passenger Cars .*** The fleet average CO_2 exhaust mass emission target values for passenger cars that are produced and delivered for sale in California each model year shall be determined as follows:

1. For passenger cars with a footprint of less than or equal to 41 square feet, the gram per mile CO_2 target value shall be selected for the appropriate model year from the following table:

Model Year	CO₂ Target Value (grams/mile)
2017	195. 0
2018	185. 0
2019	175. 0
2020	166. 0
2021	157. 0
2022	150. 0
2023	143. 0
2025	137. 0
2025 and subsequent	131. 0

2. For passenger cars with a footprint of greater than 56 square feet, the gram per mile CO_2 target value shall be selected for the appropriate model year from the following table:

Model Year	CO₂ Target Value (grams/mile)
2017	263. 0
2018	250. 0
2019	238. 0
2020	226. 0
2021	215. 0
2022	205. 0
2023	196. 0
2025	188. 0
2025 and subsequent	179. 0

3. For passenger cars with a footprint that is greater than 41 square feet and less than or equal to 56 square feet, the gram per mile CO_2 target value shall be calculated using the following equation and rounded to the nearest 0. 1 grams/mile:

$$\text{Target gCO}_2/\text{mile} = [a \times f] + b$$

Where: f is the vehicle footprint and

coefficients *a* and *b* are selected from the following table for the applicable model year.

Model year	a	b
2017	4. 53	8. 9
2018	4. 35	6. 5
2019	4. 17	4. 2
2020	4. 01	1. 9
2021	3. 84	− 0. 4
2022	3. 69	− 1. 1
2023	3. 54	− 1. 8
2024	3. 4	− 2. 5
2025 and subsequent	3. 26	− 3. 2

(**B**) Fleet Average Carbon Dioxide Target Values for Light-Duty Trucks and Medium-Duty Passenger Vehicles. The fleet average CO_2 exhaust mass emission target values for light-duty trucks and medium-duty passenger vehicles that are produced and delivered for sale in California each model year shall be determined as follows:

1. For light-duty trucks and medium-duty passenger vehicles with a footprint of less than or equal to 41 square feet, the gram per mile CO_2 target value shall be selected from the following table:

Model Year	CO_2 Target Value (grams/mile)
2017	238. 0
2018	227. 0
2019	220. 0
2020	212. 0
2021	195. 0
2022	186. 0
2023	176. 0
2025	168. 0
2025 and subsequent	159. 0

2. For light-duty trucks and medium-duty passenger vehicles with a footprint of greater than 41 square feet and less than or equal to the maximum footprint value specified in the table below for each model year, the gram/mile CO_2 target value shall be calculated using the following equation and rounded to the nearest 0. 1 grams/mile:

$$Target\ gCO_2/mile = [a \times f] + b$$

Where: f is the vehicle footprint and

coefficients a and b are selected from the following table for the applicable model year.

Model year	Maximum Footprint	a	b
2017	50. 7	4. 87	38. 3
2018	60. 2	4. 76	31. 6
2019	66. 4	4. 68	27. 7
2020	68. 3	4. 57	24. 6
2021	73. 5	4. 28	19. 8
2022	74. 0	4. 09	17. 8
2023	74. 0	3. 91	16. 0
2024	74. 0	3. 74	14. 2
2025 and subsequent	74. 0	3. 58	12. 5

3. For light-duty trucks and medium-duty passenger vehicles with a footprint that is greater than the minimum footprint value specified in the table below and less than or equal to the maximum footprint value specified in the table below for each model year, the gram/mile CO_2 target value shall be calculated using the following equation and rounded to the nearest 0. 1 grams/mile:

$$Target\ gCO_2/mile = [a \times f] + b$$

Where: f is the vehicle footprint and

coefficients a and b are selected from the following table for the applicable model year.

Model year	Minimum Footprint	Maximum Footprint	a	b
2017	50. 7	66. 0	4. 04	80. 5
2018	60. 2	66. 0	4. 04	75. 0

4. For light-duty trucks and medium-duty passenger vehicles with a footprint that is greater than the minimum value specified in the table below for each model year, the gram/mile CO_2 target value shall be selected for the applicable model year from the following table:

Model year	Minimum Footprint	CO_2 target value (grams/mile)
2017	66.0	347.0
2018	66.0	342.0
2019	66.4	339.0
2020	68.3	337.0
2021	73.5	335.0
2022	74.0	321.0
2023	74.0	306.0
2024	74.0	291.0
2025 and subsequent	74.0	277.0

(C) Calculation of Manufacturer-Specific Carbon Dioxide Fleet Average Standards . For each model year, each manufacturer must comply with fleet average CO_2 standards for passenger cars and for light-duty trucks plus medium-duty passenger vehicles, as applicable, calculated for that model year as follows. For each model year, a manufacturer must calculate separate fleet average CO_2 values for its passenger car fleet and for its combined light-duty truck plus medium-duty passenger vehicle fleet using the CO_2 target values in subsection (a)(A). These calculated CO_2 values are the manufacturer-specific fleet average CO_2 standards for passenger cars and for light-duty trucks plus medium-duty passenger vehicles, as applicable, which apply for that model year.

1. A CO_2 target value shall be calculated in accordance with subparagraph (a) (1)(A) or (a)(1)(B), as applicable, for each unique combination of model type and footprint value.

2. Each CO_2 target value, determined for each unique combination of model type and footprint value, shall be multiplied by the total production of that model type/footprint combination for the applicable model year.

3. The resulting products shall be summed, and that sum shall be divided by the total production of passenger cars or total combined production of light-duty trucks and medium-duty passenger vehicles, as applicable, in that model year. The result shall be rounded to the nearest whole gram per mile. This result shall be the applicable fleet average CO_2 standard for the manufacturer's passenger car fleet or its combined light-duty truck and medium-duty passenger vehicle fleet, as applicable.

(2) *Nitrous Oxide* (N_2O) *and Methane* (CH_4) *Exhaust Emission Standards for Passenger Cars, Light-Duty Trucks, and Medium-Duty Passenger Vehicles.* Each manufacturer's fleet of combined passenger automobile, light-duty trucks, and medium-duty passenger vehicles must comply with N_2O and CH_4 standards using either the provisions of subsection (a)(2)(A), subsection (a)(2)(B), or subsection (a)(2)(C). Except with prior approval of the Executive Officer, a manufacturer may not use the provisions of both subsection (a)(2)(A) and subsection (a)(2)(B) in the same model year. For example, a manufacturer may not use the provisions of subsection (a)(2)(A) for their passenger automobile fleet and the provisions of subsection (a)(2)(B) for their light-duty truck and medium-duty passenger vehicle fleet in the same model year. The manufacturer may use the provisions of both subsections (a)(2)(A) and (a)(2)(C) in the same model year. For example, a manufacturer may meet the N_2O standard in subsection (a) (2)(A) 1 and an alternative CH_4 standard determined under subsection (a)(2)(C).

(A) *Standards Applicable to Each Test Group.*

1. Exhaust emissions of N_2O shall not exceed 0.010 grams per mile at full useful life, as measured on the FTP (40 CFR, Part 86, Subpart B), as amended by the "California 2015 and Subsequent Model Criteria Pollutant Exhaust Emission Standards and Test Procedures and 2017 and Subsequent Model Greenhouse Gas Exhaust Emission Standards and Test Procedures for Passenger Cars, Light Duty Trucks, and Medium Duty Vehicles." Manufacturers may optionally determine an alternative N_2O standard under subsection (a)(2) (C).

2. Exhaust emissions of CH_4 shall not exceed 0.030 grams per mile at full useful life, as measured on the FTP (40 CFR, Part 86, Subpart B), as amended by the "California 2015 and Subsequent Model Criteria Pollutant Exhaust Emission Standards and Test Procedures and 2017 and Subsequent Model Greenhouse Gas Exhaust Emission Standards and Test Procedures for Passenger Cars, Light-Duty Trucks, and Medium-Duty Vehicles." Manufacturers may optionally determine an alternative CH_4 standard under subsection (a)

(2)(C).

(B) *Including N₂O and CH₄ in Fleet Averaging Program.* Manufacturers may elect to not meet the emission standards in subsection (a)(2)(A). Manufacturers making this election shall measure N_2O and CH_4 emissions for each unique combination of model type and footprint value on both the FTP test cycle and the Highway Fuel Economy test cycle at full useful life, multiply the measured N_2O emissions value by 298 and the measured CH_4 emissions value by 25, and include both of these adjusted N_2O and CH_4 full useful life values in the fleet average calculations for passenger automobiles and light-duty trucks plus medium-duty passenger vehicles, as calculated in accordance with subsection (a)(2)(A)(D).

(C) *Optional Use of Alternative N₂O and/or CH₄ Standards.* Manufacturers may select an alternative standard applicable to a test group, for either N_2O or CH_4, or both. For example, a manufacturer may choose to meet the N_2O standard in subsection (a)(2)(A)1 and an alternative CH_4 standard in lieu of the standard in subsection (a)(2)(A)2. The alternative standard for each pollutant must be less stringent than the applicable exhaust emission standard specified in subsection (a)(2)(A). Alternative N_2O and CH_4 standards apply to emissions as measured on the FTP (40 CFR, Part 86, Subpart B), as amended by the "California 2015 and Subsequent Model Criteria Pollutant Exhaust Emission Standards and Test Procedures and 2017 and Subsequent Model Greenhouse Gas Exhaust Emission Standards and Test Procedures for Passenger Cars, Light-Duty Trucks, and Medium-Duty Vehicles," for the full useful life, and become the applicable certification and in-use emission standard(s) for the test group. Manufacturers using an alternative standard for N_2O and/or CH_4 must calculate emission debits according to the provisions of subsection (a)(2)(D) for each test group/alternative standard combination. Debits must be included in the calculation of total credits or debits generated in a model year as required under subsection (b)(1)(B). Flexible fuel vehicles (or other vehicles certified for multiple fuels) must meet these alternative standards when tested on all applicable test fuel type.

(D) *CO₂-Equivalent Debits.* CO_2-equivalent debits for test groups using an alternative N_2O and/or CH_4 standard as determined under (a)(2)(C) shall be calculated according to the following equation and rounded to the nearest whole gram per mile:

$$Debits = GWP \times (Production) \times (AltStd\text{-}Std)$$

Where:

Debits = N_2O or CH_4 CO_2-equivalent debits for a test group using an alternative N_2O or

CH_4 standard;

　　GWP = 25 if calculating CH_4 debits and 298 if calculating N_2O debits;

　　Production = The number of vehicles of that test group produced and delivered for sale in California;

　　AltStd = The alternative standard (N_2O or CH_4) selected by the manufacturer under (a)(2)(C); and

　　Std = The exhaust emission standard for N_2O or CH_4 specified in (a)(2)(A).

(3) *Alternative Fleet Average Standards for Manufacturers with Limited U. S. Sales.* Manufacturers meeting the criteria in this subsection (a)(3) may request that the Executive Officer establish alternative fleet average CO_2 standards that would apply instead of the standards in subsection (a)(1).

　　(A) *Eligibility for Alternative Standards.* Eligibility as determined in this subsection (a)(3) shall be based on the total sales of combined passenger cars, light-duty trucks, and medium-duty passenger vehicles. The terms "sales" and "sold" as used in this subsection (a)(3) shall mean vehicles produced and delivered for sale (or sold) in the states and territories of the United States. For the purpose of determining eligibility the sales of related companies shall be aggregated according to the provisions of section 1900. To be eligible for alternative standards established under this subsection (a)(3), the manufacturer's average sales for the three most recent consecutive model years must remain below 5 000. If a manufacturer's average sales for the three most recent consecutive model years exceeds 4 999, the manufacturer will no longer be eligible for exemption and must meet applicable emission standards as follows.

　　　　1. If a manufacturer's average sales for three consecutive model years exceeds 4 999, and if the increase in sales is the result of corporate acquisitions, mergers, or purchase by another manufacturer, the manufacturer shall comply with the emission standards described in subsections (a)(1) and (a)(2), as applicable, beginning with the first model year after the last year of the three consecutive model years.

　　　　2. If a manufacturer's average sales for three consecutive model years exceeds 4 999 and is less than 50 000, and if the increase in sales is solely the result of the manufacturer's expansion in vehicle production (not the result of corporate acquisitions, mergers, or purchase by another manufacturer), the manufacturer shall comply with the emission standards described in subsections (a)(1) and (a)(2), as applicable, beginning with the second model year after the last year of the three consecutive model years.

(B) *Requirements for New Entrants into the U. S. Market.* New entrants are those manufacturers without a prior record of automobile sales in the United States and without prior certification to (or exemption from, under 40 CFR § 86. 1801 – 12 (k)) greenhouse gas emission standards in 40 CFR § 86. 1818 – 12 or greenhouse gas standards in section 1961. 1. In addition to the eligibility requirements stated in subsection (a)(3) (A), new entrants must meet the following requirements:

1. In addition to the information required under subsection (a)(3)(D), new entrants must provide documentation that shows a clear intent by the company to actually enter the U. S. market in the years for which alternative standards are requested. Demonstrating such intent could include providing documentation that shows the establishment of a U. S. dealer network, documentation of work underway to meet other U. S. requirements (e. g., safety standards), or other information that reasonably establishes intent to the satisfaction of the Executive Officer.

2. Sales of vehicles in the U. S. by new entrants must remain below 5 000 vehicles for the first two model years in the U. S. market and the average sales for any three consecutive years within the first five years of entering the U. S. market must remain below 5 000 vehicles. Vehicles sold in violation of these limits will be considered not covered by the certificate of conformity and the manufacturer will be subject to penalties on an individual-vehicle basis for sale of vehicles not covered by a certificate. In addition, violation of these limits will result in loss of eligibility for alternative standards until such point as the manufacturer demonstrates two consecutive model years of sales below 5 000 automobiles.

3. A manufacturer with sales in the most recent model year of less than 5 000 automobiles, but where prior model year sales were not less than 5 000 automobiles, is eligible to request alternative standards under subsection (a)(3). However, such a manufacturer will be considered a new entrant and subject to the provisions regarding new entrants in this subsection (a)(3), except that the requirement to demonstrate an intent to enter the U. S. market in subsection (a)(3)(B)(1) shall not apply.

(C) *How to Request Alternative Fleet Average Standards.* Eligible manufacturers may petition for alternative standards for up to five consecutive model years if sufficient information is available on which to base such standards.

1. To request alternative standards starting with the 2017 model year, eligible manufacturers must submit a completed application no later than July 30, 2013.

2. To request alternative standards starting with a model after 2017, eligible manufacturers must submit a completed application no later than 36 months prior to the start of the first model year to which the alternative standards would apply.

3. The application must contain all the information required in subsection (a) (3)(D), and must be signed by a chief officer of the company. If the Executive Officer determines that the content of the request is incomplete or insufficient, the manufacturer will be notified and given an additional 30 days to amend the request.

4. A manufacturer may elect to petition for alternative standards under this subsection (a)(3)(C) by submitting to ARB a copy of the data and information submitted to EPA as required under 40 CFR § 86. 1818 – 12 (g), incorporated by reference in and amended by the "California 2015 and Subsequent Model Criteria Pollutant Exhaust Emission Standards and Test Procedures and 2017 and Subsequent Model Greenhouse Gas Exhaust Emission Standards and Test Procedures for Passenger Cars, Light-Duty Trucks, and Medium-Duty Vehicles," and the EPA approval of the manufacturer's request for alternative fleet average standards for the 2017 through 2025 MY National Greenhouse Gas Program.

(**D**) *Data and Information Submittal Requirements.* Eligible manufacturers requesting alternative standards under subsection (a)(3) must submit the following information to the California Air Resources Board. The Executive Officer may request additional information as s/he deems appropriate. The completed request must be sent to the California Air Resources Board at the following address: Chief, Mobile Source Operations Division, California Air Resources Board, 9480 Telstar Avenue, Suite 4, El Monte, California 91731.

1. *Vehicle Model and Fleet Information.*

a. The model years to which the requested alternative standards would apply, limited to five consecutive model years.

b. Vehicle models and projections of production volumes for each model year.

c. Detailed description of each model, including the vehicle type, vehicle mass, power, footprint, and expected pricing.

d. The expected production cycle for each model, including new model introductions and redesign or refresh cycles.

2. *Technology Evaluation Information.*

a. The CO_2 reduction technologies employed by the manufacturer on each vehicle model, including information regarding the cost and CO_2-reducing effectiveness. Include technologies that improve air conditioning efficiency and reduce air conditioning system leakage, and any "off-cycle" technologies that potentially provide benefits outside the operation represented by the FTP and the HWFET.

b. An evaluation of comparable models from other manufacturers, including CO_2 results and air conditioning credits generated by the models. Comparable vehicles should be similar, but not necessarily identical, in the following respects: vehicle type, horsepower, mass, power-to-weight ratio, footprint, retail price, and any other relevant factors. For manufacturers requesting alternative standards starting with the 2017 model year, the analysis of comparable vehicles should include vehicles from the 2012 and 2013 model years, otherwise the analysis should at a minimum include vehicles from the most recent two model years.

c. A discussion of the CO_2 – reducing technologies employed on vehicles offered outside of the U. S. market but not available in the U. S., including a discussion as to why those vehicles and/or technologies are not being used to achieve CO_2 reductions for vehicles in the U. S. market.

d. An evaluation, at a minimum, of the technologies projected by the California Air Resources Board in the "Staff Report: Initial Statement of Reasons for Proposed Rulemaking, Public Hearing to Consider the "LEV III" Amendments to The California Greenhouse Gas and Criteria Pollutant Exhaust and Evaporative Emission Standards and Test Procedures and to the On-Board Diagnostic System Requirements for Passenger Cars, Light-Duty Trucks, and Medium-Duty Vehicles, and to the Evaporative Emission Requirements for Heavy-Duty Vehicles" and the appendices to this report, released on December 7, 2011, as those technologies likely to be used to meet greenhouse gas emission standards and the extent to which those technologies are employed or projected to be employed by the manufacturer. For any technology that is not projected to be fully employed, the manufacturer must explain why this is the case.

3. *Information Supporting Eligibility.*

a. U. S. sales for the three previous model years and projected sales for the model years for which the manufacturer is seeking alternative standards.

b. Information regarding ownership relationships with other manufacturers, including details regarding the application of the provisions of 40 CFR § 86. 1838 – 01 (b)(3) and section 1900 regarding the aggregation of sales of related companies.

(E) *Alternative Standards.* Upon receiving a complete application, the Executive Officer will review the application and determine whether an alternative standard is warranted. If the Executive Officer judges that an alternative standard is warranted, the following standards shall apply. For the purposes of this subsection (a)(3)(E), an "ultra-small volume manufacturer" shall mean a manufacturer that meets the requirements of subsection (a)(3).

1. At the beginning of the model year that is three model years prior to the model year for which an alternative standard is requested, each ultra-small volume manufacturer shall identify all vehicle models from the model year that is four model years prior to the model year for which an alternative standard is requested, certified by a large volume manufacturer that are comparable to that small volume manufacturer's vehicle models for the model year for which an alternative standard is requested, based on model type and footprint value. The ultra-small volume manufacturer shall demonstrate to the Executive Officer the appropriateness of each comparable vehicle model selected. Upon approval of the Executive Officer, s/he shall provide to the ultra-small volume manufacturer the target grams CO_2 per mile for each vehicle model type and footprint value that is approved. The ultra-small volume manufacturer shall calculate its fleet average CO_2 standard in accordance with subsection (a)(1)(C) based on these target grams CO_2 per mile values provided by the Executive Officer.

2. In the 2017 and subsequent model years, an ultra-small volume manufacturer shall either:

a. not exceed its fleet average CO_2 standard calculated in accordance with subsection (a)(1)(C) based on the target grams CO_2 per mile values provided by the Executive Officer; or

b. upon approval of the Executive Officer, if an ultra-small volume manufacturer demonstrates a vehicle model uses an engine, transmission, and emission control system and has a footprint value that are identical to a configuration certified for sale in California by a large volume manufacturer, those ultra-small volume manufacturer vehicle models are exempt from meeting the requirements in paragraph 2. a of this subsection.

(F) *Restrictions on Credit Trading.* Manufacturers subject to alternative standards approved by the Executive Officer under this subsection (a)(3) may not trade credits to another manufacturer. Transfers of credits between a manufacturer's car and truck fleets are allowed.

(4) *Greenhouse Gas Emissions Values for Electric Vehicles, "Plug-In" Hybrid Electric Vehicles, and Fuel Cell Vehicles.*

（A）*Electric Vehicle Calculations.*

1. For each unique combination of model type and footprint value, a manufacturer shall calculate the City CO_2 Value using the following formula:

$$City\ CO_2\ Value = (270\ gCO_2e/kWh) \times E_{EV} - 0.25 \times CO_{2target}$$

Where E_{EV} is measured directly from each cycle for each test vehicle of battery electric vehicle technology in units of kilowatt-hours per mile (per SAE J1634, incorporated herein by reference).

2. For each unique combination of model type and footprint value, a manufacturer shall calculate the Highway CO_2 Value using the following formula:

$$Highway\ CO_2\ Value = (270\ gCO_2e/kWh) \times E_{EV} - 0.25 \times CO_{2target}$$

Where E_{EV} is measured directly from each cycle for each test vehicle of battery electric vehicle technology in units of kilowatt-hours per mile (per SAE J1634, incorporated herein by reference).

（B）*"Plug-In" Hybrid Electric Vehicle Calculations.*

For each unique combination of model type and footprint value, a manufacturer shall calculate the City CO_2 Value and the Highway CO_2 Value using the following formulas:

$$City\ CO_2\ Value = GHG_{urban}$$

and

$$Highway\ CO_2\ Value = GHG_{highway}$$

Where GHG_{urban} and $GHG_{highway}$ are measured in accordance with section G. 12 of the "California Exhaust Emission Standards and Test Procedures for 2009 through 2017 Model Zero-Emission Vehicles and Hybrid Electric Vehicles, in the Passenger Car, Light-Duty Truck and Medium-Duty Vehicle Classes" or the "California Exhaust Emission Standards and Test Procedures for 2018 and Subsequent Model Zero-Emission Vehicles and Hybrid Electric Vehicles, in the Passenger Car, Light-Duty Truck and Medium-Duty Vehicle Classes," as applicable.

（C）*Fuel Cell Vehicle Calculations.*

For each unique combination of model type and footprint value, a manufacturer shall calculate the City CO_2 Value and the Highway CO_2 Value using the following formulas:

$$City\ CO_2 = GHG_{FCV} = (9132\ gCO_2e/kg\ H_2) \times H_{FCV} - G_{upstream}$$

and

$$Highway\ CO_2 = GHG_{FCV} = (9132\ gCO_2e/kg\ H_2) \times H_{FCV} - G_{upstream}$$

Where H_{FCV} means hydrogen consumption in kilograms of hydrogen per mile, measured for the applicable test cycle, in accordance with SAE J2572 (published October 2008), incorporated herein by reference.

(5) *Calculation of Fleet Average Carbon Dioxide Value.*

(A) For each unique combination of model type and footprint value, a manufacturer shall calculate a combined city/highway CO_2 exhaust emission value as follows:

$$0.55 \times City\ CO_2\ Value + 0.45 \times Highway\ CO_2\ Value$$

"City" CO_2 exhaust emissions shall be measured using the FTP test cycle (40 CFR, Part 86, Subpart B), as amended by the "California 2015 and Subsequent Model Criteria Pollutant Exhaust Emission Standards and Test Procedures and 2017 and Subsequent Model Greenhouse Gas Exhaust Emission Standards and Test Procedures for Passenger Cars, Light Duty Trucks, and Medium Duty Vehicles. " "Highway" CO_2 exhaust emission shall be measured using the using the Highway Fuel Economy Test (HWFET; 40 CFR 600 Subpart B).

(B) Each combined city/highway CO_2 exhaust emission, determined for each unique combination of model type and footprint value, shall be multiplied by the total production of that model type/footprint combination for the applicable model year.

(C) The resulting products shall be summed, and that sum shall be divided by the total production of passenger cars or total combined production of light-duty trucks and medium-duty passenger vehicles, as applicable, in that model year. The result shall be rounded to the nearest whole gram per mile. This result shall be the manufacturer's actual sales-weighted fleet average CO_2 value for the manufacturer's passenger car fleet or its combined light-duty truck and medium-duty passenger vehicle fleet, as applicable.

(D) For each model year, a manufacturer must demonstrate compliance with the fleet average requirements in section (a) (1) based on one of two options applicable throughout the model year, either:

Option 1: the total number of passenger cars, light-duty trucks, and medium-duty passenger vehicles that are certified to the California exhaust emission standards in section 1961.3, and are produced and delivered for sale in California; or

Option 2: the total number of passenger cars, light-duty trucks, and medium-duty passenger vehicles that are certified to the California exhaust emission standards in this section 1961.3, and are produced and delivered for sale in California, the District of Columbia, and all states that have adopted California's greenhouse gas emission standards for that model year pursuant to Section 177 of the federal Clean Air Act (42 U.S.C. § 7507).

1. A manufacturer that selects compliance Option 2 must notify the Executive Of-

ficer of that selection, in writing, prior to the start of the applicable model year or must comply with Option 1. Once a manufacturer has selected compliance Option 2, that selection applies unless the manufacturer selects Option 1 and notifies the Executive Officer of that selection in writing before the start of the applicable model year.

2. When a manufacturer is demonstrating compliance using Option 2 for a given model year, the term "in California" as used in section 1961. 3 means California, the District of Columbia, and all states that have adopted California's greenhouse gas emission standards for that model year pursuant to Section 177 of the federal Clean Air Act (42 U. S. C. § 7507).

3. A manufacturer that selects compliance Option 2 must provide to the Executive Officer separate values for the number of vehicles in each model type and footprint value produced and delivered for sale in the District of Columbia and for each individual state within the average and the City CO_2 Value and Highway CO_2 exhaust emission values that apply to each model type and footprint value.

(6) *Credits for Reduction of Air Conditioning Direct Emissions.* Manufacturers may generate A/C Direct Emissions Credits by implementing specific air conditioning system technologies designed to reduce air conditioning direct emissions over the useful life of their vehicles. A manufacturer may only use an A/C Direct Emissions Credit for vehicles within a model type upon approval of the A/C Direct Emissions Credit for that model type by the Executive Officer. The conditions and requirements for obtaining approval of an A/C Direct Emissions Credit are described in (A) through (F), below.

(A) Applications for approval of an A/C Direct Emissions Credit must be organized by model type. The applications must also include:

- vehicle make and
- number of vehicles within the model type that will be equipped with the air conditioning system to which the leakage credit shall apply.

Separate applications must be submitted for any two configurations of an A/C system with differences other than dimensional variation.

(B) To obtain approval of the A/C Direct Emissions Credit, the manufacturer must demonstrate through an engineering evaluation that the A/C system under consideration reduces A/C direct emissions. The demonstration must include all of the following elements:

- the amount of A/C Direct Emissions Credit requested, in grams of CO_2 – equivalent per mile（gCO_2e/mi）；

- the calculations identified in section（a）（6）（C）justifying that credit amount；

- schematic of the A/C system；

- specifications of the system components with sufficient detail to allow reproduction of the calculation；and

- an explanation describing what efforts have been made to minimize the number of fittings and joints and to optimize the components in order to minimize leakage.

Calculated values must be carried to at least three significant figures throughout the calculations, and the final credit value must be rounded to one tenth of a gram of CO_2 – equivalent per mile（gCO_2e/mi）.

（C）The calculation of A/C Direct Emissions Credit depends on the refrigerant or type of system, and is specified in paragraphs 1, 2, and 3 of this subsection.

1. HFC – 134a vapor compression systems

For A/C systems that use HFC – 134a refrigerant, the A/C Direct Emissions Credit is calculated using the following formula：

$$A/C \ Direct \ Credit = Direct \ Credit \ Baseline \times \left(1 - \frac{LR}{AvgLR}\right)$$

Where：

Direct Credit Baseline = 12. 6 gCO_2e/mi for passenger cars；

Direct Credit Baseline = 15. 6 gCO_2e/mi for light-duty trucks and medium-duty passenger vehicles；

Avg LR = 16. 6 grams/year for passenger cars；

Avg LR = 20. 7 grams/year for light-duty trucks and medium-duty passenger vehicles；

LR = the larger of SAE LR or Min LR；

Where：

SAE LR = initial leak rate evaluated using SAE International's Surface Vehicle Standard SAE J2727（Revised February 2012）, incorporated by reference, herein；

Min LR = 8. 3 grams/year for passenger car A/C systems with belt-driven compressors；

Min LR = 10. 4 grams/year for light-duty truck and medium-duty passenger vehicle A/C systems with belt-driven compressors；

Min LR = 4. 1 grams/year for passenger car A/C systems with electric compressors；

Min LR = 5. 2 grams/year for light-duty truck and medium-duty passenger vehicle A/C systems with electric compressors.

Note: Initial leak rate is the rate of refrigerant leakage from a newly manufactured A/C system in grams of refrigerant per year. The Executive Officer may allow a manufacturer to use an updated version of SAE J2727 or an alternate method if s/he determines that the updated SAE J2727 or the alternate method provides more accurate estimates of the initial leak rate of A/C systems than the February 2012 version of SAE J2727 does.

2. Low-GWP vapor compression systems

For A/C systems that use a refrigerant having a GWP of 150 or less, the A/C Direct Emissions Credit shall be calculated using the following formula:

$$A/C\ Direct\ Credit = Low\ GWP\ Credit\text{-}High\ Leak\ Penalty$$

Where:

$$Low\ GWP\ Credit = Max\ Low\ GWP\ Credit \times \left(1 - \frac{GWP}{1\ 430}\right)$$

and

High Leak Penalty

$$= \begin{cases} Max\ High\ Leak\ Penalty, & if\ SAE\ LR > Avg\ LR; \\ Max\ High\ Leak\ Penalty \times \dfrac{SAELR - MinLR}{AvgLR - MinLR}, & if\ Min\ LR < SAE\ LR \leq Avg\ LR; \\ 0, & if\ SAE\ LR \leq Min\ LR. \end{cases}$$

Where:

Max Low GWP Credit = 13.8 gCO_2e/mi for passenger cars;

Max Low GWP Credit = 17.2 gCO_2e/mi for light-duty trucks and medium-duty passenger vehicles;

GWP = the global warming potential of the refrigerant over a 100-year horizon, as specified in section (a)(6)(F);

Max High Leak Penalty = 1.8 gCO_2e/mi for passenger cars;

Max High Leak Penalty = 2.1 gCO_2e/mi for light-duty trucks and medium-duty passenger vehicles;

Avg LR = 13.1 g/yr for passenger cars;

Avg LR = 16.6 g/yr for light-duty trucks and medium-duty passenger vehicles;

and where:

SAE LR = initial leak rate evaluated using SAE International's Surface Vehicle Standard SAE J2727 (Revised February 2012);

Min LR = 8.3 g/yr for passenger cars;

Min LR = 10.4 g/yr for light-duty trucks and medium-duty passenger vehicles.

Note: Initial leak rate is the rate of refrigerant leakage from a newly manufactured A/C system in grams of refrigerant per year. The Executive Officer may allow a manufacturer to use an updated version of SAE J2727 or an alternate method if s/he determines that the update or the alternate method provides more accurate estimates of the initial leak rate of A/C systems than the February 2012 version of SAE J2727 does.

3. Other A/C systems

For an A/C system that uses a technology other than vapor compression cycles, an A/C Direct Emissions Credit may be approved by the Executive Officer. The amount of credit requested must be based on demonstration of the reduction of A/C direct emissions of the technology using an engineering evaluation that includes verifiable laboratory test data, and cannot exceed 13. 8 gCO_2e/mi for passenger cars and 17. 2 gCO_2e/mi for light-duty trucks and medium-duty passenger vehicles.

(**D**) The total leakage reduction credits generated by the air conditioning system shall be calculated separately for passenger cars, and for light-duty trucks and medium-duty passenger vehicles, according to the following formula:

$$Total\ Credits\ (g/mi)\ = A/C\ Direct\ Credit \times Production$$

Where:

A/C Direct Credit is calculated as specified in subsection (a)(6)(C).

Production = The total number of passenger cars or light-duty trucks plus medium-duty passenger vehicles, whichever is applicable, produced and delivered for sale in California, with the air conditioning system to which the A/D Direct Credit value from subsection (a)(6)(C) applies.

(**E**) The results of subsection (a)(6)(D), rounded to the nearest whole gram per mile, shall be included in the manufacturer's credit/debit totals calculated in subsection (b)(1)(B).

(**F**) The following values for refrigerant global warming potential (GWP), or alternative values as determined by the Executive Officer, shall be used in the calculations of this subsection (a)(6). The Executive Officer shall determine values for refrigerants not included in this subsection (a)(6)(F) upon request by a manufacturer, based on findings by the Intergovernmental Panel on Climate Change (IPCC) or from other applicable research studies.

Refrigerant	GWP
HFC – 134a	1 430
HFC – 152a	124
HFO – 1234yf	4
CO_2	1

(7) Credits for Improving Air Conditioning System Efficiency. Manufacturers may generate CO_2 credits by implementing specific air conditioning system technologies designed to reduce air conditioning-related CO_2 emissions over the useful life of their passenger cars, light-duty trucks, and/or medium-duty passenger vehicles. Credits shall be calculated according to this subsection (a)(7) for each air conditioning system that the manufacturer is using to generate CO_2 credits. The eligibility requirements specified in subsection (a) (7)(E) must be met before an air conditioning system is allowed to generate credits.

(A) Air conditioning efficiency credits are available for the following technologies in the gram per mile amounts indicated for each vehicle category in the following table:

Air Conditioning Technology	Passenger Cars (g/mi)	Light-Duty Trucks and Medium-Duty Passenger Vehicles (g/mi)
Reduced reheat, with externally-controlled, variable-displacement compressor (e.g. a compressor that controls displacement based on temperature setpoint and/or cooling demand of the air conditioning system control settings inside the passenger compartment).	1.5	2.2
Reduced reheat, with externally-controlled, fixed-displacement or pneumatic variable displacement compressor (e.g. a compressor that controls displacement based on conditions within, or internal to, the air conditioning system, such as head pressure, suction pressure, or evaporator outlet temperature).	1.0	1.4
Default to recirculated air with closed-loop control of the air supply (sensor feedback to control interior air quality) whenever the ambient temperature is 75°F or higher: Air conditioning systems that operated with closed-loop control of the air supply at different temperatures may receive credits by submitting an engineering analysis to the Administrator for approval.	1.5	2.2

续表

Air Conditioning Technology	Passenger Cars (g/mi)	Light-Duty Trucks and Medium-Duty Passenger Vehicles (g/mi)
Default to recirculated air with open-loop control air supply (no sensor feedback) whenever the ambient temperature is 75° F or higher. Air conditioning systems that operate with open-loop control of the air supply at different temperatures may receive credits by submitting an engineering analysis to the Administrator for approval.	1.0	1.4
Blower motor controls which limit wasted electrical energy (e. g. pulse width modulated power controller).	0.8	1.1
Internal heat exchanger (e. g. a device that transfers heat from the high-pressure, liquid-phase refrigerant entering the evaporator to the low-pressure, gas-phase refrigerant exiting the evaporator).	1.0	1.4
Improved condensers and/or evaporators with system analysis on the component (s) indicating a coefficient of performance improvement for the system of greater than 10% when compared to previous industry standard designs).	1.0	1.4
Oil separator. The manufacturer must submit an engineering analysis demonstrating the increased improvement of the system relative to the baseline design, where the baseline component for comparison is the version which a manufacturer most recently had in production on the same vehicle design or in a similar or related vehicle model. The characteristics of the baseline component shall be compared to the new component to demonstrate the improvement.	0.5	0.7

(B) Air conditioning efficiency credits are determined on an air conditioning system basis. For each air conditioning system that is eligible for a credit based on the use of one or more of the items listed in subsection (a) (7) (A), the total credit value is the sum of the gram per mile values listed in subsection (a) (7) (A) for each item that applies to the air conditioning system. However, the total credit value for an air conditioning system may not be greater than 5. 0 grams per mile for any passenger car or 7. 2 grams per mile for any light-duty truck or medium-duty passenger vehicle.

(C) The total efficiency credits generated by an air conditioning system shall be calculated separately for passenger cars and for light-duty trucks plus medium-duty passenger vehicles according to the following formula:

$$Total\ Credits\ (g/mi)\ = Credit \times Production$$

Where:

Credit = the CO_2 efficiency credit value in grams per mile determined in subsection (a)(7)(B) or (a)(7)(E), whichever is applicable.

Production = The total number of passenger cars or light-duty trucks plus medium-duty passenger vehicles, whichever is applicable, produced and delivered for sale in California, with the air conditioning system to which to the efficiency credit value from subsection (a)(7)(B) applies.

(D) The results of subsection (a)(7)(C), rounded to the nearest whole gram per mile, shall be included in the manufacturer's credit/debit totals calculated in subsection (b)(1)(B).

(E) For the purposes of this subsection (a)(7)(E), the AC17 Test Procedure shall mean the AC17 Air Conditioning Efficiency Test Procedure set forth in 40 CFR § 86.167 – 17, incorporated in and amended by the "California 2015 and Subsequent Model Criteria Pollutant Exhaust Emission Standards and Test Procedures and 2017 and Subsequent Model Greenhouse Gas Exhaust Emission Standards and Test Procedures for Passenger Cars, Light-Duty Trucks, and Medium-Duty Vehicles."

1. For each air conditioning system selected by the manufacturer to generate air conditioning efficiency credits, the manufacturer shall perform the AC17 Test Procedure.

2. Using good engineering judgment, the manufacturer must select the vehicle configuration to be tested that is expected to result in the greatest increased CO_2 emissions as a result of the operation of the air conditioning system for which efficiency credits are being sought. If the air conditioning system is being installed in passenger cars, light-duty trucks, and medium-duty passenger vehicles, a separate determination of the quantity of credits for passenger cars and for light-duty trucks and medium-duty passenger vehicles must be made, but only one test vehicle is required to represent the air conditioning system, provided it represents the worst-case impact of the system on CO_2 emissions.

3. For each air conditioning system selected by the manufacturer to generate air conditioning efficiency credits, the manufacturer shall perform the AC17 Test Procedure according to the following requirements. Each air conditioning system shall be tested as follows:

a. Perform the AC17 test on a vehicle that incorporates the air conditioning system with the credit-generating technologies.

b. Perform the AC17 test on a vehicle which does not incorporate the credit-generating technologies. The tested vehicle must be similar to the vehicle tested under subsection (a)(7)(E)(3) a.

c. Subtract the CO_2 emissions determined from testing under subsection (a)(7) (E)(3) a from the CO_2 emissions determined from testing under subsection (a)(7)(E) (3) b and round to the nearest 0.1 grams/mile. If the result is less than or equal to zero, the air conditioning system is not eligible to generate credits. If the result is greater than or equal to the total of the gram per mile credits determined under subsection (a)(7)(B), then the air conditioning system is eligible to generate the maximum allowable value determined under subsection (a)(7)(B). If the result is greater than zero but less than the total of the gram per mile credits determined under subsection (a)(7)(B), then the air conditioning system is eligible to generate credits in the amount determined by subtracting the CO_2 emissions determined from testing under subsection (a)(7)(E)(3) a from the CO_2 emissions determined from testing under subsection (a)(7)(E)(3) b and rounding to the nearest 0.1 grams/mile.

4. For the first model year for which an air conditioning system is expected to generate credits, the manufacturer must select for testing the highest-selling subconfiguration within each vehicle platform that uses the air conditioning system. Credits may continue to be generated by the air conditioning system installed in a vehicle platform provided that:

a. The air conditioning system components and/or control strategies do not change in any way that could be expected to cause a change in its efficiency;

b. The vehicle platform does not change in design such that the changes could be expected to cause a change in the efficiency of the air conditioning system; and

c. The manufacturer continues to test at least one sub-configuration within each platform using the air conditioning system, in each model year, until all sub-configurations within each platform have been tested.

5. Each air conditioning system must be tested and must meet the testing criteria in order to be allowed to generate credits. Using good engineering judgment, in the first model year for which an air conditioning system is expected to generate credits, the manufacturer must select for testing the highest-selling subconfiguration within each vehicle platform using the air conditioning system. Credits may continue to be generated by an air conditioning system in subsequent model years if the manufacturer continues to test at least one sub-configuration within each platform on annually, as long as the air conditioning system and vehicle platform do not change substantially.

(8) *Off-Cycle Credits.* Manufacturers may generate credits for CO_2-reducing technologies where the CO_2 reduction benefit of the technology is not adequately captured on the FTP and/or the HWFET. These technologies must have a measurable, demonstrable, and verifiable real-world CO_2 reduction that occurs outside the conditions of the FTP and the HWFET. These optional credits are referred to as "off-cycle" credits. Off-cycle technologies used to generate emission credits are considered emission-related components subject to applicable requirements, and must be demonstrated to be effective for the full useful life of the vehicle. Unless the manufacturer demonstrates that the technology is not subject to in-use deterioration, the manufacturer must account for the deterioration in their analysis. The manufacturer must use one of the three options specified in this subsection (a)(8) to determine the CO_2 gram per mile credit applicable to an off-cycle technology. The manufacturer should notify the Executive Officer in its pre-model year report of its intention to generate any credits under this subsection (a)(8).

(A) *Credit available for certain off-cycle technologies.*

1. The manufacturer may generate a CO_2 gram/mile credit for certain technologies as specified in the following table, provided that each technology is applied to the minimum percentage of the manufacturer's total U. S. production of passenger cars, light-duty trucks, and medium-duty passenger vehicles specified in the table in each model year for which credit is claimed. Technology definitions are in subsection (e).

Off-Cycle Technology	*Passenger Cars (g/mi)*	*Light-Duty Trucks and Medium-Duty Passenger Vehicles (g/mi)*	*Minimum Total Percent of U. S. Production*
Active aerodynamics	0. 6	1. 0	10
High efficiency exterior lighting	1. 1	1. 1	10
Engine heat recovery	0. 7 per 100W of capacity	0. 7 per 100W of capacity	10
Engine start-stop (idle-off)	2. 9	4. 5	10
Active transmission warm-up	1. 8	1. 8	10

续表

Off-Cycle Technology	Passenger Cars (g/mi)	Light-Duty Trucks and Medium-Duty Passenger Vehicles (g/mi)	Minimum Total Percent of U. S. Production
Active engine warm-up	1.8	1.8	10
Electric heater circulation pump	1.0	1.5	n/a
Solar roof panels	3.0	3.0	n/a
Thermal control	≤3.0	≤4.3	n/a

a. Credits may also be accrued for thermal control technologies as defined in sub-section (e) in the amounts shown in the following table:

Thermal Control Technology	Credit value: Passenger Cars (g/mi)	Credit Value: Light-Duty Trucks and Medium-Duty Passenger Vehicles (g/mi)
Glass or glazing	≤2.9	≤3.9
Active seat ventilation	1.0	1.3
Solar reflective paint	0.4	0.5
Passive cabin ventilation	1.7	2.3
Active cabin ventilation	2.1	2.8

b. The maximum credit allowed for thermal control technologies is limited to 3.0 g/mi for passenger cars and to 4.3 g/mi for light-duty trucks and medium-duty passenger vehicles. The maximum credit allowed for glass or glazing is limited to 2.9 g/mi for passenger cars and to 3.9 g/mi for light-duty trucks and medium-duty passenger vehicles.

c. Glass or glazing credits are calculated using the following equation:

$$Credit = \left[Z \times \sum_{i=1}^{n} \frac{T_i \times G_i}{G} \right]$$

Where:

Credit = the total glass or glazing credits, in grams per mile, for a vehicle, which may not exceed 3.0 g/mi for passenger cars or 4.3 g/mi for light-duty trucks and medium-duty passenger vehicles;

Z = 0.3 for passenger cars and 0.4 for light-duty trucks and medium-duty passenger vehicles;

G_i = the measured glass area of window i, in square meters and rounded to the nearest tenth;

G = the total glass area of the vehicle, in square meters and rounded to the nearest tenth;

T_i = the estimated temperature reduction for the glass area of window i, determined using the following formula:

$$T_i = 0.3987 \times (Tts_{base} - Tts_{new})$$

Where:

Tts_{new} = the total solar transmittance of the glass, measured according to ISO 13837: 2008, "Safety glazing materials-Method for determination of solar transmittance" (incorporated by reference, herein).

Tts_{base} = 62 for the windshield, side-front, side-rear, rear-quarter, and backlite locations, and 40 for rooflite locations.

2. The maximum allowable decrease in the manufacturer's combined passenger car and light-duty truck plus medium-duty passenger vehicle fleet average CO_2 emissions attributable to use of the default credit values in subsection (a)(8)(A)1 is 10 grams per mile. If the total of the CO_2 g/mi credit values from the table in subsection (a)(8)(A)1 does not exceed 10 g/mi for any passenger automobile or light truck in a manufacturer's fleet, then the total off-cycle credits may be calculated according to subsection (a)(8) (D). If the total of the CO_2 g/mi credit values from the table in subsection (a)(8)(A)1 exceeds 10 g/mi for any passenger car, light-duty truck, or medium-duty passenger vehicle in a manufacturer's fleet, then the gram per mile decrease for the combined passenger car and light-duty truck plus medium-duty passenger vehicle fleet must be determined according to subsection (a)(8)(A) 2. a to determine whether the 10 g/mi limitation has been exceeded.

a. Determine the gram per mile decrease for the combined passenger car and light-duty truck plus medium-duty passenger vehicle fleet using the following formula:

$$Decrease = \frac{Credits \times 1,000,000}{[(Prod_c \times 195,264) + (Prod_T \times 225,865)]}$$

Where:

Credits = The total of passenger car and light-duty truck plus medium-duty passenger vehicles credits, in Megagrams, determined according to subsection (a)(8)(D) and limited to those credits accrued by using the default gram per mile values in subsection (a) (8)(A) 1.

$Prod_C$ = The number of passenger cars produced by the manufacturer and delivered for sale in the U. S.

$Prod_T$ = The number of light-duty trucks and medium-duty passenger vehicles produced by the manufacturer and delivered for sale in the U. S.

b. If the value determined in subsection (a)(8)(A) 2. a is greater than 10 grams per mile, the total credits, in Megagrams, that may be accrued by a manufacturer u-sing the default gram per mile values in subsection (a)(8)(A) 1 shall be determined u-sing the following formula:

$$Credit \ (Megagrams) \ = \ \frac{[10 \times ((Prod_c \times 195\ 264) + (Prod_T \times 225\ 865))]}{1\ 000\ 000}$$

Where:

$Prod_C$ = The number of passenger cars produced by the manufacturer and delivered for sale in the U. S. = The number of passenger cars produced by the manufacturer and deliv-ered for sale in the U. S.

$Prod_T$ = The number of light-duty trucks and medium-duty passenger vehicles produced by the manufacturer and delivered for sale in the U. S.

c. If the value determined in subsection (a)(8)(A) 2. a is not greater than 10 grams per mile, then the credits that may be accrued by a manufacturer using the default gram per mile values in subsection (a)(8)(A) 1 do not exceed the allowable limit, and total credits may be determined for each category of vehicles according to subsection (a) (8)(D).

d. If the value determined in subsection (a)(8)(A) 2. a is greater than 10 grams per mile, then the combined passenger car and light-duty truck plus medium-duty passenger vehicle credits, in Megagrams, that may be accrued using the calculations in sub-section (a)(8)(D) must not exceed the value determined in subsection (a)(8)(A) 2. b. This limitation should generally be done by reducing the amount of credits attributable to the vehicle category that caused the limit to be exceeded such that the total value does not exceed the value determined in subsection (a)(8)(A) 2. b.

3. In lieu of using the default gram per mile values specified in subsection (a) (8)(A) 1 for specific technologies, a manufacturer may determine an alternative value for any of the specified technologies. An alternative value must be determined using one of the methods specified in subsection (a)(8)(B) or subsection (a)(8)(C).

(B) *Technology demonstration using EPA 5-cycle methodology* . To demonstrate an off-cycle technology and to determine a CO_2 credit using the EPA 5-cycle methodology, the manufacturer shall determine the off-cycle city/highway combined carbon-related exhaust emissions benefit by using the EPA 5-cycle methodology described in 40 CFR Part 600. Testing shall be performed on a representative vehicle, selected using good engineering judgment, for each model type for which the credit is being demonstrated. The emission benefit of a technology is determined by testing both with and without the off-cycle technology operating. Multiple off-cycle technologies may be demonstrated on a test vehicle. The manufacturer shall conduct the following steps and submit all test data to the Executive Officer.

1. Testing without the off-cycle technology installed and/or operating. Determine carbon-related exhaust emissions over the FTP, the HWFET, the US06, the SC03, and the cold temperature FTP test procedures according to the test procedure provisions specified in 40 CFR part 600 subpart B and using the calculation procedures specified in §600. 113 – 08 of this chapter. Run each of these tests a minimum of three times without the off-cycle technology installed and operating and average the per phase (bag) results for each test procedure. Calculate the 5-cycle weighted city/highway combined carbon-related exhaust emissions from the averaged per phase results, where the 5-cycle city value is weighted 55% and the 5-cycle highway value is weighted 45%. The resulting combined city/highway value is the baseline 5-cycle carbon-related exhaust emission value for the vehicle.

2. Testing with the off-cycle technology installed and/or operating. Determine carbon-related exhaust emissions over the US06, the SC03, and the cold temperature FTP test procedures according to the test procedure provisions specified in 40 CFR part 600 subpart B and using the calculation procedures specified in 40 CFR §600. 113 – 08. Run each of these tests a minimum of three times with the off-cycle technology installed and operating and average the per phase (bag) results for each test procedure. Calculate the 5-cycle weighted city/highway combined carbon-related exhaust emissions from the averaged per phase results, where the 5-cycle city value is weighted 55% and the 5-cycle highway value is weighted 45%. Use the averaged per phase results for the FTP and HWFET determined in subsection (a)(8)(B) 1 for operation without the off-cycle technology in this calculation. The resulting combined city/highway value is the 5-cycle carbon-related exhaust emission value showing the off-cycle benefit of the technology but excluding any benefit of the

technology on the FTP and HWFET.

3. Subtract the combined city/highway value determined in subsection （a）（8）（B） 1 from the value determined in subsection （a）（8）（B） 2. The result is the off-cycle benefit of the technology or technologies being evaluated. If this benefit is greater than or equal to three percent of the value determined in subsection （a）（8）（B） 1 then the manufacturer may use this value, rounded to the nearest tenth of a gram per mile, to determine credits under subsection （a）（8）（C）.

4. If the value calculated in subsection （a）（8）（B） 3 is less than two percent of the value determined in subsection （a）（8）（B） 1, then the manufacturer must repeat the testing required under subsections （a）（8）（B） 1 and （a）（8）（B） 2, except instead of running each test three times they shall run each test two additional times. The off-cycle benefit of the technology or technologies being evaluated shall be calculated as in subsection （a）（8）（B） 3 using all the tests conducted under subsections （a）（8）（B） 1, （a）（8）（B） 2, and （a）（8）（B） 4. If the value calculated in subsection （a）（8）（B） 3 is less than two percent of the value determined in subsection （a）（8）（B） 1, then the manufacturer must verify the emission reduction potential of the off-cycle technology or technologies using the EPA Vehicle Simulation Tool, and if the results support a credit value that is less than two percent of the value determined in subsection （a）（8）（B） 1 then the manufacturer may use the off-cycle benefit of the technology or technologies calculated as in subsection （a）（8）（B） 3 using all the tests conducted under subsections （a）（8）（B） 1, （a）（8）（B） 2, and （a）（8）（B） 4, rounded to the nearest tenth of a gram per mile, to determine credits under subsection （a）（8）（C）.

（C） *Review and approval process for off-cycle credits.*

1. *Initial steps required.*

a. A manufacturer requesting off-cycle credits under the provisions of subsection （a）（8）（B） must conduct the testing and/or simulation described in that paragraph.

b. A manufacturer requesting off-cycle credits under subsection （a）（8）（B） must conduct testing and/or prepare engineering analyses that demonstrate the in-use durability of the technology for the full useful life of the vehicle.

2. *Data and information requirements.* The manufacturer seeking off-cycle credits must submit an application for off-cycle credits determined under subsection （a）（8）（B）. The application must contain the following:

a. A detailed description of the off-cycle technology and how it functions to reduce CO_2 emissions under conditions not represented on the FTP and HWFET.

b. A list of the vehicle model (s) which will be equipped with the technology.

c. A detailed description of the test vehicles selected and an engineering analysis that supports the selection of those vehicles for testing.

d. All testing and/or simulation data required under subsection (a)(8)(B), as applicable, plus any other data the manufacturer has considered in the analysis.

e. An estimate of the off-cycle benefit by vehicle model and the fleetwide benefit based on projected sales of vehicle models equipped with the technology.

f. An engineering analysis and/or component durability testing data or whole vehicle testing data demonstrating the in-use durability of the off-cycle technology components.

3. *Review of the off-cycle credit application.* Upon receipt of an application from a manufacturer, the Executive Officer will do the following:

a. Review the application for completeness and notify the manufacturer within 30 days if additional information is required.

b. Review the data and information provided in the application to determine if the application supports the level of credits estimated by the manufacturer.

4. *Decision on off-cycle application.* The Executive Officer will notify the manufacturer in writing of its decision to approve or deny the application within 60 days of receiving a complete application, and if denied, the Executive Officer will provide the reasons for the denial.

(D) *Calculation of total off-cycle credits.* Total off-cycle credits in grams per mile of CO_2 (rounded to the nearest tenth of a gram per mile) shall be calculated separately for passenger cars and light-duty trucks plus medium-duty passenger vehicles according to the following formula:

$$Total\ Credits\ (g/mi) = Credit \times Production$$

Where:

Credit = the credit value in grams per mile determined in subsection (a)(8)(A) or subsection (a)(8)(B).

Production = The total number of passenger cars or light-duty trucks plus medium – duty passenger vehicles, whichever is applicable, produced and delivered for sale in California, produced with the off-cycle technology to which to the credit value determined in subsection (a)(8)(A) or subsection (a)(8)(B) applies.

(9) *Credits for certain full-size pickup trucks.* Full-size pickup trucks may be eligible for additional credits based on the implementation of hybrid technologies or on exhaust emission performance, as described in this subsection (a)(9). Credits may be generated under either subsection (a)(9)(A) or subsection (a)(9)(B) for a qualifying pickup truck, but not both.

(A) *Credits for implementation of gasoline-electric hybrid technology.* Full-size pickup trucks that implement hybrid gasoline-electric technologies may be eligible for an additional credit under this subsection (a)(9)(A). Pickup trucks using the credits under this subsection (a)(9)(A) may not use the credits described in subsection (a)(9)(B).

1. Full-size pickup trucks that are mild hybrid gasoline-electric vehicles and that are produced in the 2017 through 2021 model years are eligible for a credit of 10 grams/mile. To receive this credit, the manufacturer must produce a quantity of mild hybrid full-size pickup trucks such that the proportion of production of such vehicles, when compared to the manufacturer's total production of full-size pickup trucks, is not less than the amount specified in the table below for each model year.

Model year	*Required minimum percent of full-size pickup trucks*
2017	30%
2018	40%
2019	55%
2020	70%
2021	80%

2. Full-size pickup trucks that are strong hybrid gasoline-electric vehicles and that are produced in the 2017 through 2025 model years are eligible for a credit of 20 grams/mile. To receive this credit, the manufacturer must produce a quantity of strong hybrid full-size pickup trucks such that the proportion of production of such vehicles, when compared to the manufacturer's total production of full-size pickup trucks, is not less than 10 percent for each model year.

(B) *Credits for emission reduction performance.* 2017 through 2021 model year full-size pickup trucks that achieve carbon-related exhaust emission values below the applicable target value determined in subsection (a)(1)(B) may be eligible for an additional credit. Pickup trucks using the credits under this subsection (a)(9)(B) may not use the credits described in subsection (a)(9)(A).

1. Full-size pickup trucks that achieve carbon-related exhaust emissions less than or equal to the applicable target value determined in subsection (a)(1)(B) multiplied by 0.85 (rounded to the nearest gram per mile) and greater than the applicable target value determined in subsection (a)(1)(B) multiplied by 0.80 (rounded to the nearest gram per mile) in a model year are eligible for a credit of 10 grams/mile. A pickup truck that qualifies for this credit in a model year may claim this credit for subsequent model years through the 2021 model year if the carbon-related exhaust emissions of that pickup truck do not increase relative to the emissions in the model year in which the pickup truck qualified for the credit. To qualify for this credit in each model year, the manufacturer must produce a quantity of full-size pickup trucks that meet the emission requirements of this subsection (a)(9)(B) 1 such that the proportion of production of such vehicles, when compared to the manufacturer's total production of full-size pickup trucks, is not less than the amount specified in the table below for each model year.

Model year	Required minimum percent of full-size pickup trucks
2017	15%
2018	20%
2019	28%
2020	35%
2021	40%

2. Full-size pickup trucks that achieve carbon-related exhaust emissions less than or equal to the applicable target value determined in subsection (a)(1)(B) multiplied by 0.80 (rounded to the nearest gram per mile) in a model year are eligible for a credit of 20 grams/mile. A pickup truck that qualifies for this credit in a model year may claim this credit for a maximum of five subsequent model years if the carbon-related exhaust emissions of that pickup truck do not increase relative to the emissions in the model year in which the pickup truck first qualified for the credit. This credit may not be claimed in any model year after 2025. To qualify for this credit, the manufacturer must produce a quantity of full-size pickup trucks that meet the emission requirements of subsection (a)(9)(B) 1 such that the proportion of production of such vehicles, when compared to the manufacturer's total production of full-size pickup trucks, is not less than 10 percent in each model year.

(C) *Calculation of total full-size pickup truck credits.* Total credits in grams per mile of CO_2 (rounded to the nearest whole gram per mile) shall be calculated for qualifying full-size pickup trucks according to the following formula:

$$Total\ Credits\ (g/mi) = (10 \times Production_{10}) + (20 \times Production_{20})$$

Where:

$Production_{10}$ = The total number of full-size pickup trucks produced and delivered for sale in California with a credit value of 10 grams per mile from subsection (a)(9)(A) and subsection (a)(9)(B).

$Production_{20}$ = The total number of full-size pickup trucks produced and delivered for sale in California with a credit value of 20 grams per mile from subsection (a)(9)(A) and subsection (a)(9)(B).

(10) *Greenhouse Gas In-Use Compliance Standards.* The in-use exhaust CO_2 emission standard shall be the combined city/highway exhaust emission value calculated according to the provisions of subsection (a)(5)(A) for the vehicle model type and footprint value multiplied by 1.1 and rounded to the nearest whole gram per mile. For vehicles that are capable of operating on multiple fuels, a separate value shall be determined for each fuel that the vehicle is capable of operating on. These standards apply to in-use testing performed by the manufacturer pursuant to the "California 2015 and Subsequent Model Criteria Pollutant Exhaust Emission Standards and Test Procedures and 2017 and Subsequent Model Greenhouse Gas Exhaust Emission Standards and Test Procedures for Passenger Cars, Light-Duty Trucks, and Medium-Duty Vehicles."

(11) *Mid-Term Review of the 2022 through 2025 MY Standards.* The Executive Officer shall conduct a mid-term review to re-evaluate the state of vehicle technology to determine whether any adjustments to the stringency of the 2022 through 2025 model year standards are appropriate. California's mid-term review will be coordinated with its planned full participation in EPA's mid-term evaluation as set forth in 40 CFR § 86.1818 – 12 (h).

(b) *Calculation of Greenhouse Gas Credits/Debits.* Credits that are earned as part of the 2012 through 2016 MY National greenhouse gas program shall not be applicable to California's greenhouse gas program. Debits that are earned as part of the 2012 through 2016 MY National greenhouse gas program shall not be applicable to California's greenhouse gas program.

(1) Calculation of Greenhouse Gas Credits for Passenger Cars, Light-Duty Trucks, and Medium-Duty Passenger Vehicles.

(A) A manufacturer that achieves fleet average CO_2 values lower than the fleet average CO values lower than the fleet average CO_2 requirement for the corresponding model year shall receive credits for each model year in units of g/mi. A manufacturer that achieves fleet average CO_2 values higher than the fleet average CO_2 requirement for the corresponding model year shall receive debits for each model year in units of g/mi. Manufacturers must calculate greenhouse gas credits and greenhouse gas debits separately for passenger cars and for combined light-duty trucks and medium-duty passenger vehicles as follows:

CO_2 Credits or Debits = (CO_2 Standard-Manufacturer's Fleet Average CO_2 Value) × (Total No. of Vehicles Produced and Delivered for Sale in California, Including ZEVs and HEVs).

Where:

CO_2 Standard = the applicable standard for the model year as determined in subsection (a)(1)(C);

Manufacturer's Fleet Average CO_2 Value = average calculated according to subsection (a)(5);

(B) A manufacturer's total Greenhouse Gas credits or debits generated in a model year shall be the sum of its CO_2 credits or debits and any of the following credits or debits, if applicable. The manufacturer shall calculate, maintain, and report Greenhouse Gas credits or debits separately for its passenger car fleet and for its light-duty truck plus medium-duty passenger vehicle fleet.

1. Air conditioning leakage credits earned according to the provisions of subsection (a)(6);

2. Air conditioning efficiency credits earned according to the provisions of subsection (a)(7);

3. Off-cycle technology credits earned according to the provisions of subsection (a)(8).

4. CO_2-equivalent debits earned according to the provisions of subsection (a)(2)(D).

(2) A manufacturer with 2017 and subsequent model year fleet average Greenhouse Gas values greater than the fleet average CO_2 standard applicable for the corresponding model year shall receive debits in units of g/mi Greenhouse Gas equal to the amount of negative credits determined by the aforementioned equation. For the 2017 and subsequent model years, the total g/mi Greenhouse Gas credits or debits earned for passenger cars and for light-duty trucks and medium-duty passenger vehicles shall be summed together. The resulting amount shall constitute the g/mi Greenhouse Gas credits or debits accrued by the manufacturer for the model year.

(3) *Procedure for Offsetting Greenhouse Gas Debits.*

(A) A manufacturer shall equalize Greenhouse Gas emission debits by earning g/mi Greenhouse Gas emission credits in an amount equal to the g/mi Greenhouse Gas debits, or by submitting a commensurate amount of g/mi Greenhouse Gas credits to the Executive Officer that were earned previously or acquired from another manufacturer. A manufacturer shall equalize combined Greenhouse Gas debits for passenger cars, light-duty trucks, and medium-duty passenger vehicles within five model years after they are earned. If emission debits are not equalized within the specified time period, the manufacturer shall be subject to the Health and Safety Code section 43211 civil penalty applicable to a manufacturer which sells a new motor vehicle that does not meet the applicable emission standards adopted by the state board. The cause of action shall be deemed to accrue when the emission debits are not equalized by the end of the specified time period. For a manufacturer demonstrating compliance under Option 2 in subsection (a)(5)(D), the emission debits that are subject to a civil penalty under Health and Safety Code section 43211 shall be calculated separately for California, the District of Columbia, and each individual state that is included in the fleet average greenhouse gas requirements in subsection (a)(1). These emission debits shall be calculated for each individual state using the formula in subsections (b)(1) and (b)(2), except that the "Total No. of Vehicles Produced and Delivered for Sale in California, including ZEVs and HEVs" shall be calculated separately for the District of Columbia and each individual state.

For the purposes of Health and Safety Code section 43 211, the number of passenger cars not meeting the state board's emission standards shall be determined by dividing the total amount of g/mi Greenhouse Gas emission debits for the model year calculated for California by the g/mi Greenhouse Gas fleet average requirement for passenger car applicable for the model year in which the debits were first incurred. For the purposes of Health and Safety Code section 43 211, the number of light-duty trucks and medium-duty passenger vehicles not meeting the state board's emission standards shall be determined by dividing the total amount of g/mi Greenhouse Gas emission debits for the model year calculated for California by the g/mi Greenhouse Gas fleet average requirement for light-duty trucks and medium-duty passenger vehicles, applicable for the model year in which the debits were first incurred.

(B) Greenhouse Gas emission credits earned in the 2017 and subsequent model years shall retain full value through the fifth model year after they are earned, and will have no value if not used by the beginning of the sixth model year after being earned.

(4) *Use of Greenhouse Gas Emission Credits to Offset a Manufacturer's ZEV Obligations.*

(A) For a given model year, a manufacturer that has Greenhouse Gas credits remaining after equalizing all of its Greenhouse Gas debits may use those Greenhouse Gas credits to comply with its ZEV obligations for that model year, in accordance with the provisions set forth in the "California Exhaust Emission Standards and Test Procedures for 2018 and Subsequent Model Zero-Emission Vehicles and Hybrid Electric Vehicles, in the Passenger Car, Light-Duty Truck and Medium-Duty Vehicle Classes," incorporated by reference in section 1962. 2.

(B) Any Greenhouse Gas credits used by a manufacturer to comply with its ZEV obligations shall retain no value for the purposes of complying with this section 1961. 3.

(5) Credits and debits that are earned as part of the 2012 through 2016 MY National Greenhouse Gas Program, shall have no value for the purpose of complying with this section 1961. 3.

(c) *Optional Compliance with the 2017 through 2025 MY National Greenhouse Gas Program.*

For the 2017 through 2025 model years, a manufacturer may elect to demonstrate compliance with this section 1961. 3 by demonstrating compliance with the 2017 through 2025 MY National greenhouse gas program as follows:

(1) A manufacturer that selects compliance with this option must notify the Executive Officer of that selection, in writing, prior to the start of the applicable model year or must comply with 1961. 3 (a) and (b);

(2) The manufacturer must submit to ARB all data that it submits to EPA in accordance with the reporting requirements as required under 40 CFR § 86. 1865 – 12, incorporated by reference in and amended by the "California 2015 and Subsequent Model Criteria Pollutant Exhaust Emission Standards and Test Procedures and 2017 and Subsequent Model Greenhouse Gas Exhaust Emission Standards and Test Procedures for Passenger Cars, Light-

Duty Trucks, and Medium-Duty Vehicles, " for demonstrating compliance with the 2017 through 2025 MY National greenhouse gas program and the EPA determination of compliance. All such data must be submitted within 30 days of receipt of the EPA determination of compliance for each model year that a manufacturer selects compliance with this option;

(3) The manufacturer must provide to the Executive Officer separate values for the number of vehicles in each model type and footprint value produced and delivered for sale in California, the District of Columbia, and each individual state that has adopted California's greenhouse gas emission standards for that model year pursuant to Section 177 of the federal Clean Air Act (42 U. S. C. § 7507), the applicable fleet average CO_2 standards for each of these model types and footprint values, the calculated fleet average CO_2 value for each of these model types and footprint values, and all values used in calculating the fleet average CO_2 values.

(d) *Test Procedures.* The certification requirements and test procedures for determining compliance with the emission standards in this section are set forth in the "California 2015 and Subsequent Model Criteria Pollutant Exhaust Emission Standards and Test Procedures and 2017 and Subsequent Model Greenhouse Gas Exhaust Emission Standards and Test Procedures for Passenger Cars, Light-Duty Trucks, and Medium-Duty Vehicles, " incorporated by reference in section 1961. 2. In the case of hybrid electric vehicles, the certification requirements and test procedures for determining compliance with the emission standards in this section are set forth in the "California Exhaust Emission Standards and Test Procedures for 2009 through 2017 Model Zero-Emission Vehicles and Hybrid Electric Vehicles, in the Passenger Car, Light-Duty Truck and Medium-Duty Vehicle Classes, " incorporated by reference in section 1962. 1, or the "California Exhaust Emission Standards and Test Procedures for 2018 and Subsequent Model Zero-Emission Vehicles and Hybrid E-lectric Vehicles, in the Passenger Car, Light-Duty Truck and Medium-Duty Vehicle Classes, " incorporated by reference in section 1962. 2, as applicable.

(e) *Abbreviations.* The following abbreviations are used in this section 1961. 3:

"CFR" means Code of Federal Regulations.

"CH_4" means methane.

"CO_2" means carbon dioxide.

"FTP" means Federal Test Procedure.

"GHG" means greenhouse gas.

"g/mi" means grams per mile.

"GVW" means gross vehicle weight.

"GVWR" means gross vehicle weight rating.

"GWP" means the global warming potential.

"HEV" means hybrid-electric vehicle.

"HWFET" means Highway Fuel Economy Test (HWFET; 40 CFR 600 Subpart B).

"LDT" means light-duty truck.

"LVW" means loaded vehicle weight.

"MDPV" means medium-duty passenger vehicle.

"mg/mi" means milligrams per mile.

"MY" means model year.

"NHTSA" means National Highway Traffic Safety Administration.

"N_2O" means nitrous oxide.

"ZEV" means zero-emission vehicle.

(**f**) *Definitions Specific to this Section*. The following definitions apply to this section 1961. 3:

(1) "A/C Direct Emissions" means any refrigerant released from a motor vehicle's air conditioning system.

(2) "Active Aerodynamic Improvements" means technologies that are activated only at certain speeds to improve aerodynamic efficiency by a minimum of three percent, while preserving other vehicle attributes or functions.

(3) "Active Cabin Ventilation" means devices that mechanically move heated air from the cabin interior to the exterior of the vehicle.

(4) "Active Transmission Warmup" means a system that uses waste heat from the exhaust system to warm the transmission fluid to an operating temperature range quickly using a heat exchanger in the exhaust system, increasing the overall transmission efficiency by reducing parasitic losses associated with the transmission fluid, such as losses related to friction and fluid viscosity.

(5) "Active Engine Warmup" means a system using waste heat from the exhaust system to warm up targeted parts of the engine so that it reduces engine friction losses and enables the closed-loop fuel control to activate more quickly. It allows a faster transition from cold operation to warm operation, decreasing CO_2 emissions.

(6) "Active Seat Ventilation" means a device that draws air from the seating surface which is in contact with the occupant and exhausts it to a location away from the seat.

（7） "Blower motor controls which limit waste energy" means a method of controlling fan and blower speeds that does not use resistive elements to decrease the voltage supplied to the motor.

（8） "Default to recirculated air mode" means that the default position of the mechanism which controls the source of air supplied to the air conditioning system shall change from outside air to recirculated air when the operator or the automatic climate control system has engaged the air conditioning system （i. e. , evaporator is removing heat）, except under those conditions where dehumidification is required for visibility （i. e. , defogger mode）. In vehicles equipped with interior air quality sensors （e. g. , humidity sensor, or carbon dioxide sensor）, the controls may determine proper blend of air supply sources to maintain freshness of the cabin air and prevent fogging of windows while continuing to maximize the use of recirculated air. At any time, the vehicle operator may manually select the non-recirculated air setting during vehicle operation but the system must default to recirculated air mode on subsequent vehicle operations （i. e. , next vehicle start）. The climate control system may delay switching to recirculation mode until the interior air temperature is less than the outside air temperature, at which time the system must switch to recirculated air mode.

（9） "Electric Heater Circulation Pump" means a pump system installed in a stop-start equipped vehicle or in a hybrid electric vehicle or plug-in hybrid electric vehicle that continues to circulate hot coolant through the heater core when the engine is stopped during a stop-start event. This system must be calibrated to keep the engine off for 1 minute or more when the external ambient temperature is 30 deg F.

（10） "Emergency Vehicle" means a motor vehicle manufactured primarily for use as an ambulance or combination ambulance-hearse or for use by the United States Government or a State or local government for law enforcement.

（11） "Engine Heat Recovery" means a system that captures heat that would otherwise be lost through the exhaust system or through the radiator and converting that heat to electrical energy that is used to meet the electrical requirements of the vehicle. Such a system must have a capacity of at least 100W to achieve 0. 7 g/mi of credit. Every additional 100W of capacity will result in an additional 0. 7 g/mi of credit.

（12） "Engine Start-Stop" means a technology which enables a vehicle to automatically turn off the engine when the vehicle comes to a rest and restart the engine when the driver applies pressure to the accelerator or releases the brake.

（13） "EPA Vehicle Simulation Tool" means the "EPA Vehicle Simulation Tool" as incorporated by reference in 40 CFR §86.1 in the Notice of Proposed Rulemaking for EPA's 2017 and subsequent MY National Greenhouse Gas Program, as proposed at 76 Fed. Reg. 74854, 75357 (December 1, 2011).

(14) "Executive Officer" means the Executive Officer of the California Air Resources Board.

(15) "Footprint" means the product of average track width (rounded to the nearest tenth of an inch) and wheelbase (measured in inches and rounded to the nearest tenth of an inch), divided by 144 and then rounded to the nearest tenth of a square foot, where the average track width is the average of the front and rear track widths, where each is measured in inches and rounded to the nearest tenth of an inch.

(16) "Federal Test Procedure" or "FTP" means 40 CFR, Part 86, Subpart B, as amended by the "California 2015 and Subsequent Model Criteria Pollutant Exhaust Emission Standards and Test Procedures and 2017 and Subsequent Model Greenhouse Gas Exhaust Emission Standards and Test Procedures for Passenger Cars, Light-Duty Trucks, and Medium-Duty Vehicles."

(17) "Full-size pickup truck" means a light-duty truck that has a passenger compartment and an open cargo box and which meets the following specifications:

1. A minimum cargo bed width between the wheelhouses of 48 inches, measured as the minimum lateral distance between the limiting interferences (pass-through) of the wheelhouses. The measurement shall exclude the transitional arc, local protrusions, and depressions or pockets, if present. An open cargo box means a vehicle where the cargo box does not have a permanent roof or cover. Vehicles produced with detachable covers are considered "open" for the purposes of these criteria.

2. A minimum open cargo box length of 60 inches, where the length is defined by the lesser of the pickup bed length at the top of the body and the pickup bed length at the floor, where the length at the top of the body is defined as the longitudinal distance from the inside front of the pickup bed to the inside of the closed endgate as measured at the height of the top of the open pickup bed along vehicle centerline, and the length at the floor is defined as the longitudinal distance from the inside front of the pickup bed to the inside of the closed endgate as measured at the cargo floor surface along vehicle centerline.

3. A minimum towing capability of 5 000 pounds, where minimum towing capability is determined by subtracting the gross vehicle weight rating from the gross combined weight rating, or a minimum payload capability of 1 700 pounds, where minimum payload capability is determined by subtracting the curb weight from the gross vehicle weight rating.

(**18**) "Greenhouse Gas" means the following gases: carbon dioxide, methane, nitrous oxide, and hydrofluorocarbons.

(**19**) "GWP" means the global warming potential of the refrigerant over a 100-year horizon, as specified in Intergovernmental Panel on Climate Change (IPCC) 2007: Climate Change 2007 – The Physical Science Basis. S. Solomon et al. (editors), Contribution of Working Group I to the Fourth Assessment Report of the Intergovernmental Panel on Climate Change, Cambridge University Press, Cambridge, UK and New York, NY, USA, ISBN 0 – 521 – 70596 – 7, or determined by ARB if such information is not available in the IPCC Fourth Assessment Report.

(**20**) "High Efficiency Exterior Lighting" means a lighting technology that, when installed on the vehicle, is expected to reduce the total electrical demand of the exterior lighting system by a minimum of 60 watts when compared to conventional lighting systems. To be eligible for this credit the high efficiency lighting must be installed in the following components: parking/position, front and rear turn signals, front and rear side markers, stop/brake lights (including the center-mounted location), taillights, backup/reverse lights, and license plate lighting.

(**21**) "Improved condensers and/or evaporators" means that the coefficient of performance (COP) of air conditioning system using improved evaporator and condenser designs is 10 percent higher, as determined using the bench test procedures described in SAE J2765 "Procedure for Measuring System COP of a Mobile Air Conditioning System on a Test Bench," when compared to a system using standard, or prior model year, component designs. SAE J2765 is incorporated by reference herein. The manufacturer must submit an engineering analysis demonstrating the increased improvement of the system relative to the baseline design, where the baseline component(s) for comparison is the version which a manufacturer most recently had in production on the same vehicle design or in a similar or related vehicle model. The dimensional characteristics(e. g. , tube configuration/thickness/spacing, and fin density) of the baseline component(s) shall be compared to the new component(s) to demonstrate the improvement in coefficient of performance.

(**22**) "Mild hybrid gasoline-electric vehicle" means a vehicle that has start/stop capability and regenerative braking capability, where the recaptured braking energy over the FTP is at least 15 percent but less than 75 percent of the total braking energy, where the percent of recaptured braking energy is measured and calculated according to 40 CFR

§ 600. 108 (g).

(23) "Model Type" means a unique combination of car line, basic engine, and transmission class.

(24) "2012 through 2016 MY National Greenhouse Gas Program" means the national program that applies to new 2012 through 2016 model year passenger cars, light-duty-trucks, and medium-duty passenger vehicles as adopted by the U. S. Environmental Protection Agency on April 1, 2010 (75 Fed. Reg. 25324, 25677 (May 7, 2010)).

(25) "2017 through 2025 MY National Greenhouse Gas Program" means the national program that applies to new 2017 through 2025 model year passenger cars, light-duty-trucks, and medium-duty passenger vehicles as adopted by the U. S. Environmental Protection Agency as codified in 40 CFR Part 86, Subpart S.

(26) "Oil separator" means a mechanism that removes at least 50 percent of the oil entrained in the oil/refrigerant mixture exiting the compressor and returns it to the compressor housing or compressor inlet, or a compressor design that does not rely on the circulation of an oil/refrigerant mixture for lubrication.

(27) "Passive Cabin Ventilation" means ducts or devices which utilize convective airflow to move heated air from the cabin interior to the exterior of the vehicle.

(28) "Plug-in Hybrid Electric Vehicle" means "off-vehicle charge capable hybrid electric vehicle" as defined in the "California Exhaust Emission Standards and Test Procedures for 2018 and Subsequent Model Zero-Emission Vehicles and Hybrid Electric Vehicles, in the Passenger Car, Light-Duty Truck and Medium-Duty Vehicle Classes. "

(29) "Reduced reheat, with externally controlled, fixed-displacement or pneumatic variable displacement compressor" means a system in which the output of either compressor is controlled by cycling the compressor clutch off-and-on via an electronic signal, based on input from sensors (e. g. , position or setpoint of interior temperature control, interior temperature, evaporator outlet air temperature, or refrigerant temperature) and air temperature at the outlet of the evaporator can be controlled to a level at 41°F, or higher.

(30) "Reduced reheat, with externally-controlled, variable displacement compressor" means a system in which compressor displacement is controlled via an electronic signal, based on input from sensors (e. g. , position or setpoint of interior temperature control, interior temperature, evaporator outlet air temperature, or refrigerant temperature) and air temperature at the outlet of the evaporator can be controlled to a level at 41°F, or higher.

(**31**) "SC03" means the SC03 test cycle as set forth in the "California 2015 and Subsequent Model Criteria Pollutant Exhaust Emission Standards and Test Procedures and 2017 and Subsequent Model Greenhouse Gas Exhaust Emission Standards and Test Procedures for Passenger Cars, Light Duty Trucks, and Medium Duty Vehicles. "

(**32**) "Solar Reflective Paint" means a vehicle paint or surface coating which reflects at least 65 percent of the impinging infrared solar energy, as determined using ASTM standards E903 – 96 (Standard Test Method for Solar Absorptance, Reflectance, and Transmittance of Materials Using Integrating Spheres, DOI: 10. 1520/E0903 – 96 (Withdrawn 2005)), E1918 – 06 (Standard Test Method for Measuring Solar Reflectance of Horizontal and Low-Sloped Surfaces in the Field, DOI: 10. 1520/E1918 – 06), or C1549 – 09 (Standard Test Method for Determination of Solar Reflectance Near Ambient Temperature Using a Portable Solar Reflectometer, DOI: 10. 1520/C1549 – 09). These ASTM standards are incorporated by reference, herein.

(**33**) "Solar Roof Panels" means the installation of solar panels on an electric vehicle or a plug-in hybrid electric vehicle such that the solar energy is used to provide energy to the electric drive system of the vehicle by charging the battery or directly providing power to the electric motor with the equivalent of at least 50 Watts of rated electricity output.

(**34**) "Strong hybrid gasoline-electric vehicle" means a vehicle that has start/stop capability and regenerative braking capability, where the recaptured braking energy over the Federal Test Procedure is at least 75 percent of the total braking energy, where the percent of recaptured braking energy is measured and calculated according to 40 CFR §600. 108 (g).

(**35**) "Subconfiguration" means a unique combination within a vehicle configuration of equivalent test weight, road load horsepower, and any other operational characteristics or parameters which is accepted by USEPA.

(**36**) "US06" means the US06 test cycle as set forth in the "California 2015 and Subsequent Model Criteria Pollutant Exhaust Emission Standards and Test Procedures and 2017 and Subsequent Model Greenhouse Gas Exhaust Emission Standards and Test Procedures for Passenger Cars, Light Duty Trucks, and Medium Duty Vehicles. "

(**37**) "Worst-Case" means the vehicle configuration within each test group that is expected to have the highest CO_2 – equivalent value, as calculated in section (a)(5).

(**g**) *Severability* . Each provision of this section is severable, and in the event that any provision of this section is held to be invalid, the remainder of both this section and this article remains in full force and effect.

Note：Authority cited：Sections 39500，39600，39601，43013，43018，43018.5，43101，43104 and 43105，Health and Safety Code. Reference：Sections 39002，39003，39667，43000，43009.5，43013，43018，43018.5，43100，43101，43101.5，43102，43104，43105，43106，43204，43205 and 43211，Health and Safety Code.

§ 1976   Standards and Test Procedures for Motor Vehicle Fuel Evaporative Emissions.

(a) *[Fuel evaporative emissions from 1970 through 1977 model passenger cars and light-duty trucks ; not set forth]*

(b) (1) Evaporative emissions for 1978 and subsequent model gasoline fueled, 1983 and subsequent model liquified petroleum gas fueled, and 1993 and subsequent model alcohol fueled motor vehicles and hybrid electric vehicles subject to exhaust emission standards under this article, except petroleum fueled diesel vehicles, compressed natural gas fueled vehicles, hybrid electric vehicles that have sealed fuel systems which can be demonstrated to have no evaporative emissions, and motorcycles, shall not exceed the following standards.

(G) For 2015 and subsequent model motor vehicles, the following evaporative e-mission requirements apply:

1. A. manufacturer must certify all vehicles subject to this section to the emission standards specified in either Option 1 or Option 2 below.

a. Option 1. The evaporative emissions from 2015 and subsequent model motor vehicles, tested in accordance with the test procedure sequence described in the "California Evaporative Emission Standards and Test Procedures for 2001 and Subsequent Model Motor Vehicles," incorporated by reference in section 1976 (c), shall not exceed:

vehicle Type	*Hydrocarbon[1] Emission Standards[2]*		
	Running Loss *(grams per mile)*	*Three-Day Diurnal + Hot Soak and Two-Day Diurnal + Hot Soak*	
		Whole Vehicle *(grams per test)*	*Fuel Only[3]* *(grams per test)*
Passenger cars	0. 05	0. 350	0. 0
Light-duty trucks 6 000 lbs. GVWR and under	0. 05	0. 500	0. 0
Light-duty trucks 6 001 – 8 500 lbs. GVWR	0. 05	0. 750	0. 0
Medium-duty passenger vehicles	0. 05	0. 750	0. 0

vehicle Type	Hydrocarbon[1] Emission Standards[2]		
	Running Loss (grams per mile)	Three-Day Diurnal + Hot Soak and Two-Day Diurnal + Hot Soak	
		Whole Vehicle (grams per test)	Fuel Only[3] (grams per test)
Medium-duty vehicles (8 501 – 14 000lbs. GVWR)	0. 05	0. 750	0. 0
Heavy-duty vehicles (over 14 000 lbs, GVWR)	0. 05	0. 750	0. 0

[1] Organic Material Hydrocarbon Equivalent for alcohol-fueled vehicles.

[2] For all vehicles certified to these standards, the "useful life" shall be 15 years or 150 000 miles, whichever occurs first. Approval of vehicles that are not exhaust emission tested using a chassis dynamometer pursuant to section 1961, title 13, California Code of Regulations shall be based on an engineering evaluation of the system and data submitted by the applicant.

[3] In lieu of demonstrating compliance with the fuel-only emission standard (0. 0 grams per test) over the three-day and two-day diurnal plus hot soak tests, a manufacturer may, with advance Executive Officer approval, demonstrate compliance through an alternate test plan.

b. Option 2 . The evaporative emissions from 2015 and subsequent model motor vehicles, tested in accordance with the test procedure sequence described in the "California Evaporative Emission Standards and Test Procedures for 2001 and Subsequent Model Motor Vehicles," incorporated by reference in section 1976(c), shall not exceed:

Vehicle Type	Hydrocarbon[1] Emission Standards[2]		
	Running Loss (grams per mile)	Highest Whole Vehicle Diurnal + Hot Soak [3,4,5] (grams per test)	Canister Bleed [6] (grams per test)
Passenger cars; and Light-duty trucks 6 000 lbs. GVWR and under, and 0 – 3 750 lbs. LVW	0. 05	0. 300	0. 020
Light-duty trucks 6 000 lbs. GVWR and under, and 3 751 – 5 750 lbs. LVW	0. 05	0. 400	0. 020
Light-duty trucks 6, 001 – 8 500 lbs. GVWR; and Medium-duty passenger vehicles	0. 05	0. 500	0. 020
Medium-duty vehicles (8 501 – 14 000 lbs. GVWR); and Heavy-duty vehicles (over 14 000 lbs. GVWR)	0. 05	0. 600	0. 030

[1] Organic Material Hydrocarbon Equivalent for alcohol-fueled vehicles.

[2] Except as provided below, for all vehicles certified to these standards, the "useful life" shall be 15 years or 150 000 miles, whichever occurs first. For 2016 and previous model vehicles, 2017 and previous model vehicles >6 000 lbs. GVWR, and 2021 and previous model vehicles certified by a small volume manufacturer, the canister bleed standards are certification standards only. Manufacturers are not required to establish deterioration factors for canister bleed emissions. Approval of vehicles that are not exhaust emission tested using a chassis dynamometer pursuant to section 1961, title 13, California Code of Regulations shall be based on an engineering evaluation of the system and data submitted by the applicant.

[3] The manufacturer shall determine compliance by selecting the highest whole vehicle diurnal plus hot soak emission value of the Three-Day Diurnal Plus Hot Soak Test and of the Two-Day Diurnal Plus Hot Soak Test.

[4] *Fleet-Average Option for the Highest Whole Vehicle Diurnal Plus Hot Soak Emission Standard Within Each Emission Standard Category.* A manufacturer may optionally comply with the highest whole vehicle diurnal plus hot soak emission standards by using fleet-average hydrocarbon emission values. To participate, a manufacturer must utilize the fleet-average option for all of its emission standard categories and calculate a separate fleet-average hydrocarbon emission value for each emission standard category. The emission standard categories are as follows: (1) passenger cars and light-duty trucks 6 000 pounds GVWR and under, and 0 – 3 750 pounds LVW; (2) light-duty trucks 6 000 pounds GVWR and under, and 3 751 – 5 750 pounds LVW; (3) light-duty trucks 6 001 – 8 500 pounds GVWR and medium-duty passenger vehicles; and (4) medium-duty and heavy-duty vehicles. The fleet-average hydrocarbon emission value for each emission standard category shall be calculated as follows:

$$\sum_{i=1}^{n} [\,(number\ of\ vehicles\ in\ the\ evaporative\ family)_i \times (family\ emission\ limit)_i\,] \div \sum_{i=1}^{n} (number\ of\ vehicles\ in\ the\ evaporative\ family)_i$$

where "n" = a manufacturer's total number of Option 2 certification evaporative families within an emission standard category for a given model year;

"number of vehicles in the evaporative family" = the number of vehicles produced and delivered for sale in California in the evaporative family;

"family emission limit" = the numerical value selected by the manufacturer for the evaporative family that serves as the emission standard for the evaporative family with respect to all testing, instead of the emission standard specified in this section 1976 (b)(1)(G) 1. b. The family emission limit shall not exceed 0. 500 grams per test for passenger cars; 0. 650 grams per test for light duty trucks 6 000 pounds GVWR and under; 0. 900 grams per test for light-duty trucks 6 001 – 8 500 pounds GVWR; and 1. 000 grams for medium-duty passenger vehicles, medium-duty vehicles, and heavy-duty vehicles. In addition, the family emission limit shall be set in increments of 0. 025 grams per test.

[5] *Calculation of Hydrocarbon Credits or Debits for the Fleet-Average Option.*

(1) *Calculation of Hydrocarbon Credits or Debits.* For each emission standard category in the model year, a manufacturer shall calculate the hydrocarbon credits or debits, as follows:

[(Applicable Hydrocarbon Emission Standard for the Emission Standard Category) – (Manufacturer's Fleet-Average Hydrocarbon Emission Value for the Emission Standard Category)] × (Total Number of Affected Vehicles)

where "Total Number of Affected Vehicles" = the total number of vehicles in the evaporative families participating in the fleet-average option, which are produced and delivered for sale in California, for the emission standard category of the given model year.

A negative number constitutes hydrocarbon debits, and a positive number constitutes hydrocarbon credits accrued by the manufacturer for the given model year. Hydrocarbon credits earned in a given model year shall retain full value through the fifth model year after they are earned. At the beginning of the sixth model year, the hydrocarbon credits will have no value.

(2) *Procedure for Offsetting Hydrocarbon Debits.* A manufacturer shall offset hydrocarbon debits with hydro-

carbon credits for each emission standard category within three model years after the debits have been incurred. If total hydrocarbon debits are not equalized within three model years after they have been incurred, the manufacturer shall be subject to the Health and Safety Code section 43211 civil penalties applicable to a manufacturer which sells a new motor vehicle that does not meet the applicable emission standards adopted by the state board. The cause of action shall be deemed to accrue when the hydrocarbon debits are not equalized by the end of the specified time period. For the purposes of Health and Safety Code section 43211, the number of vehicles not meeting the state board's emission standards shall be determined by dividing the total amount of hydrocarbon debits for the model year in the emission standard category by the applicable hydrocarbon emission standard for the model year in which the debits were first incurred.

Additionally, to equalize the hydrocarbon debits that remain at the end of the three model year offset period: (1) hydrocarbon credits may be exchanged between passenger cars and light-duty trucks 6 000 pounds GVWR and under and 0 – 3 750 pounds LVW, and light-duty trucks 6 000 pounds GVWR and under and 3 751 – 5 750 pounds LVW and (2) hydrocarbon credits may be exchanged between light-duty trucks 6 001 – 8 500 pounds GVWR and medium-duty passenger vehicles, and medium-duty vehicles and heavy-duty vehicles.

[6] *Vehicle Canister Bleed Emission.* Compliance with the canister bleed emission standard shall be determined based on the Bleed Emission Test Procedure described in the "California Evaporative Emission Standards and Test Procedures for 2001 and Subsequent Model Motor Vehicles," incorporated by reference in section 1976(c), and demonstrated on a stabilized canister system. Vehicles with a non-integrated refueling canister-only system are exempt from the canister bleed emission standard.

2. *Phase-In Schedule.*

For each model year, a manufacturer shall certify, at a minimum, the specified percentage of its vehicle fleet to the evaporative emission standards set forth in section 1976(b)(1)(G)1. a. or section 1976(b)(1)(G)1. b., according to the schedule set forth below. For the purpose of this section 1976(b)(1)(G)2., the manufacturer's vehicle fleet consists of the vehicles produced and delivered for sale by the manufacturer in California that are subject to the emission standards in section 1976(b)(1)(G)1. All 2015 through 2022 model motor vehicles that are not subject to these standards pursuant to the phase-in schedule shall comply with the requirements for 2004 through 2014 model motor vehicles, as described in section 1976(b)(1)(F).

Model Years	Minimum Percentage of Vehicle Fleet[1,2]
2015, 2016, and 2017	Average of vehicles certified to section 1976(b)(1)(E) in model years 2012, 2013, and 2014[3,4]
2018 and 2019	60
2020 and 2021	80
2022 and subsequent	100

[1] For the 2018 through 2022 model years only, a manufacturer may use an alternate phase-in schedule to comply with the phase-in requirements. An alternate phase-in schedule must achieve equivalent compliance

volume by the end of the last model year of the scheduled phase-in (2022). The compliance volume is the number calculated by multiplying the percent of vehicles (based on the vehicles produced and delivered for sale by the manufacturer in California) meeting the new requirements in each model year by the number of years implemented prior to and including the last model year of the scheduled phase-in, then summing these yearly results to determine a cumulative total. The cumulative total of the five year (60/60/80/80/100) scheduled phase-in set forth above is calculated as follows: (60 × 5 years) + (60 × 4 years) + (80 × 3 years) + (80 × 2 years) + (100 × 1 year) = 1 040. Accordingly, the required cumulative total for any alternate phase-in schedule of these emission standards is 1 040. The Executive Officer shall consider acceptable any alternate phase-in schedule that results in an equal or larger cumulative total by the end of the last model year of the scheduled phase-in (2022).

[2] Small volume manufacturers are not required to comply with the phase-in schedule set forth in this table. Instead, they shall certify 100 percent of their 2022 and subsequent model year vehicle fleet to the evaporative emission standards set forth in section 1976(b)(1)(G)1. a. or section 1976(b)(1)(G) 1. b.

[3] The percentage of vehicle fleet averaged across the 2015, 2016, and 2017 model years shall be used to determine compliance with this requirement.

[4] The minimum percentage required in the 2015, 2016, and 2017 model years is determined by averaging the percentage of vehicles certified to the emission standards in section 1976(b)(1)(E) in each of the manufacturer's 2012, 2013, and 2014 model year vehicle fleets. For the purpose of calculating this average, a manufacturer shall use the percentage of vehicles produced and delivered for sale in California for the 2012, 2013, and 2014 model years. A manufacturer may calculate this average percentage using the projected sales for these model years in lieu of actual sales.

3. Carry-Over of 2014 Model-Year Evaporative Families Certified to the Zero-Fuel Evaporative Emission Standards.

A manufacturer may carry over 2014 model motor vehicles certified to the zero-fuel (0. 0 grams per test) evaporative emission standards set forth in section 1976 (b)(1)(E) through the 2019 model year and be considered compliant with the requirements of section 1976 (b)(1)(G) 1. For all motor vehicles that are certified via this carry-over provision, the emission standards set forth in section 1976 (b)(1)(E) shall apply when determining in-use compliance throughout the vehicle's useful life. If the manufacturer chooses to participate in the fleet-average option for the highest whole vehicle diurnal plus hot soak emission standard, the following family emission limits are assigned to these evaporative families for the calculation of the manufacturer's fleet-average hydrocarbon emission value.

Vehicle Type	Highest Whole Vehicle Diurnal + Hot Soak (grams per test)
Passenger cars	0. 300
Light-duty trucks 6 000 lbs. GVWR and under, and 0 – 3 750 lbs. LVW	0. 300
Light-duty trucks 6 000 lbs. GVWR and under, and 3 750 – 5 750 lbs. LVW	0. 400
Light-duty trucks 6 000 – 8 500 lbs. GVWR	0. 500

4. Pooling Provision . The following pooling provision applies to the fleet-average option for the Highest Whole Vehicle Diurnal Plus Hot Soak Emission Standard in section 1976(b)(1)(G) 1. b. and to the phase-in requirements in section 1976 (b)(1)(G) 2.

a. For the fleet-average option set forth in section 1976 (b)(1)(G) 1. b. , a manufacturer must demonstrate compliance, for each model year, based on one of two options applicable throughout the model year, either:

Pooling Option 1: the total number of passenger cars, light-duty trucks, medium-duty passenger vehicles, medium-duty vehicles, and heavy-duty vehicles that are certified to the California evaporative emission standards in section 1976 (b)(1)(G) 1. b. , and are produced and delivered for sale in California; or

Pooling Option 2: the total number of passenger cars, light-duty trucks, medium-duty passenger vehicles, medium-duty vehicles, and heavy-duty vehicles that are certified to the California evaporative emission standards in section 1976 (b)(1)(G) 1. b. , and are produced and delivered for sale in California, the District of Columbia, and all states that have adopted California's evaporative emission standards set forth in section 1976 (b) (1)(G) 1. for that model year pursuant to section 177 of the federal Clean Air Act (42 U. S. C. § 7507).

b. For the phase-in requirements in section 1976 (b)(1)(G)2. , a manufacturer must demonstrate compliance, for each model year, based on one of two options applicable throughout the model year, either:

Pooling Option 1: the total number of passenger cars, light-duty trucks, medium-duty passenger vehicles, medium-duty vehicles, and heavy-duty vehicles that are certified to the California evaporative emission standards in section 1976 (b) (1) (G) 1. , and are produced and delivered for sale in California; or

Pooling Option 2: the total number of passenger cars, light-duty trucks, medium-duty passenger vehicles, medium-duty vehicles, and heavy-duty vehicles that are certified to the California evaporative emission standards in section 1976 (b) (1) (G) 1. , and are produced and delivered for sale in California, the District of Columbia, and all states that have adopted California's evaporative emission standards set forth in section 1976 (b) (1) (G) 1. for that model year pursuant to section 177 of the federal Clean Air Act (42 U. S. C. § 7507).

c. A manufacturer that selects Pooling Option 2 must notify the Executive Officer of that selection in writing before the start of the applicable model year or must comply with Pooling Option 1. Once a manufacturer has selected Pooling Option 2, that selection applies unless the manufacturer selects Option 1 and notifies the Executive Officer of that selection in writing before the start of the applicable model year.

d. When a manufacturer is demonstrating compliance using Pooling Option 2 for a given model year, the term "in California" as used in section 1976 (b) (1) (G) means California, the District of Columbia, and all states that have adopted California's evaporative emission standards for that model year pursuant to Section 177 of the federal Clean Air Act (42 U. S. C. § 7507).

e. A manufacturer that selects Pooling Option 2 must provide to the Executive Officer separate values for the number of vehicles in each evaporative family produced and delivered for sale in the District of Columbia and for each individual state within the average.

5. Optional Certification for 2014 Model Motor Vehicles. A manufacturer may optionally certify its 2014 model motor vehicles to the evaporative emission standards set forth in section 1976(b)(1)(G)1.

(b) (2) *[Evaporative emissions standards for gasoline-fueled motorcycles; not set forth]*

(c) The test procedures for determining compliance with the standards in subsection (b) above applicable to 1978 through 2000 model year vehicles are set forth in "California Evaporative Emission Standards and Test Procedures for 1978 – 2000 Model Motor Vehicles," adopted by the state board on April 16, 1975, as last amended August 5, 1999, which is incorporated herein by reference. The test procedures for determining compliance with standards applicable to 2001 and subsequent model year vehicles are set forth in the

"California Evaporative Emission Standards and Test Procedures for 2001 and Subsequent Model Motor Vehicles," adopted by the state board on August 5, 1999, and as last amended December 6, 2012, which is incorporated herein by reference.

(d) [*Applies to motorcycles only ; not set forth*]

(e) [*Applies to motorcycles only ; not set forth*]

(f) Definition Specific to this Section.

(1) [*Applies to motorcycles only ; not set forth*]

(2) For the purpose of this section, "ultra-small volume manufacturer" means any vehicle manufacturer with California sales less than or equal to 300 new vehicles per model year based on the average number of vehicles sold by the manufacturer in the previous three consecutive model years, and "small volume manufacturer" means, for 1978 through 2000 model years, any vehicle manufacturer with California sales less than or equal to 3000 new vehicles per model year based on the average number of vehicles sold by the manufacturer in the previous three consecutive model years. For 2001 and subsequent model motor vehicles, "small volume manufacturer" has the meaning set forth in section 1900 (a).

(3) "Non-integrated refueling emission control system" is defined in 40 Code of Federal Regulations § 86. 1803 – 01.

(4) "Non-integrated refueling canister-only system" means a subclass of a non-integrated refueling emission control system, where other non-refueling related evaporative emissions from the vehicle are stored in the fuel tank, instead of in a vapor storage unit (s).

Note: Authority cited: Sections 39500, 39600, 39601, 39667, 43013, 43018, 43101, 43104, 43105, 43106 and 43107, Health and Safety Code. Reference: Section 39002, 39003, 39500, 39667, 43000, 43009.5, 43013, 43018, 43100, 43101, 43101.5, 43102, 43105, 43106, 43107, 43204 and 43205 Health and Safety Code.

后　　记

　　本书内容所涉及的政策研究和翻译工作在中国清洁发展机制基金管理中心焦小平副主任的指导下完成，研究发展部温刚、许明珠、李文杰承担了主要工作，孙玉清、王宁、谢飞、夏颖哲、傅平、田晨参加了讨论。

　　在调研和翻译工作过程中，得到了许多机构和个人的大力支持，在此一并致谢：国家发展与改革委员会应对气候变化司协助联系赴加州环保署空气资源委员会，为我中心赴美现场调研提供了帮助；中国投资协会能源发展研究中心张杰理事长和比亚迪股份有限公司从事新能源汽车研发和销售的有关人员为信息和文献收集提供了帮助，张杰理事长还在相关政策研究中参与了讨论；在我中心实习的北京外国语大学张宓、外交学院郭永鹏、北京大学余梦露、美国哥伦比亚大学郑洋等同学参与了资料整理和翻译校对工作。在加州零排放车法规和低排放车法规的译校过程中，清华大学汽车工程系的郝瀚博士提供了专业指导和帮助，特别予以感谢。

　　本书内容于我国而言是一个新的工作领域，尚无可供借鉴的材料，因此编译工作中难免存在错误和疏漏之处，敬请读者指正。

中国清洁发展机制基金管理中心

2015 年 11 月